The Proceedings of New Initiatives on

Lepton Flavor Violation and Neutrino Oscillation with High Intense Muon and Neutrino Sources

Honolulu, Hawaii   2 – 6 October 2000

The Proceedings of New Initiatives on

Lepton Flavor Violation and Neutrino Oscillation with High Intense Muon and Neutrino Sources

Editors

■ **Yoshitaka Kuno** *(Osaka University, Japan)*

■ **William R Molzon** *(University of California, Irvine, USA)*

■ **Sandip Pakvasa** *(University of Hawaii, USA)*

World Scientific
New Jersey • London • Singapore • Hong Kong

Published by

World Scientific Publishing Co. Pte. Ltd.

P O Box 128, Farrer Road, Singapore 912805

USA office: Suite 1B, 1060 Main Street, River Edge, NJ 07661

UK office: 57 Shelton Street, Covent Garden, London WC2H 9HE

British Library Cataloguing-in-Publication Data
A catalogue record for this book is available from the British Library.

ISBN 981-238-084-1

This book is printed on acid-free paper.

Printed in Singapore by Mainland Press

PREFACE

The area of physics using muons and neutrinos has become quite exciting in the particle physics. Using their high intensity sources, the physics of muons and neutrinos undertakes, in various ways, the extensive searches for new physics beyond the Standard Model, such as tests of supersymmetric grand unification (SUSY-GUT), the precision measurements of the muon and neutrino properties, and in future will extend to ambitious studies such as the determination of the 3-generation neutrino mixing matrix elements and CP violation in the lepton sector. The physics of this field is getting rich together with potential improvements of the sources. Many R&D projects, such as of a high-intensity low-energy muon source or a neutrino factory, are being carried out or being planned at various places. Some of those topics are included in the present workshop.

The international workshop on "New Initiatives of Lepton Flavor Violation and Neutrino Oscillation with Very Intense Muon and Neutrino Sources" was held from October 2-6, 2000. at the East-West Center of the University of Hawaii, Hawaii. The workshop program covers various theoretical and experimental topics on the muon and neutrino physics. The workshop consisted of 37 talks. It was attended by a total of about 60 participants from various countries around the world. The sessions were packed and the discussions were ample. The success of the workshop attributed to all the speakers for their excellent talks and to all the participants for contributing to a stimulating atmosphere.

There are many people contributing to the success of the workshop. We would like to thank the participants for their active participation, which made the workshop so enjoyable. We are especially grateful to Jan Bruce (of the University of Hawaii, High Energy Physics Group) for the excellent organization which made the workshop successful.

Yoshitaka Kuno
William Molzon
Sandip Pakvasa

Contents

MUON APPLIED SCIENCE
STATUS AT THE END OF THE 20TH CENTURY

K. NAGAMINE

Meson Science Laboratory,
Institute of Materials Structure Science, KEK
Oho 1-1, Tsukuba, Ibaraki 305-0801, Japan
E-mail:kanetada.nagamine@kek.jp
and
Muon Science Laboratory,
RIKEN (Institute of Physical and Chemical Research)
Hirosawa 2-1, Wako, Saitama 351-0198, Japan
E-mail: nagamine@postman.riken.go.jp

Potential applications of muons to various fields of scientific research have been recognized since the discovery of the muon among cosmic-rays. Recent progress is reviewed here with special emphasis on (1) muon-catalyzed fusion for fusion energy development, (2) an electron labelling method via muon spin-relaxation measurements for life science applications and (3) the application of high-energy muons and neutrinos to disaster prevention, such as the construction of a data-base for predictions of volcanic eruption and earthquakes.

1 Introduction

First, an overview is presented of the existing sources of various types of muons and their applications. As summarized in Fig.1, so far, two types of muons are available, namely, accelerator-produced muons and cosmic-ray muons. The former involves high intensity and low energy with a thin stopping range, while the later involves low intensity and high energy with a very thick stopping-range. Various interesting scientific studies have been realized by using these two types of muon beams. Accelerator-produced muons, after stopping in a mm∼cm thick target material, are used to conduct condensed-matter studies by a muon spin rotation/relaxation/resonance (μSR) method (mostly μ^+), muon-catalyzed fusion(μCF) studies (μ^-) and non-destructive element analysis studies via muonic X-ray for, e.g., biomedical applications (μ^-). Recently, owing to an ultra-slow positive muon technique, sub-μm thick materials can now be an objective of μSR studies. On the other hand, it is now known that cosmic-ray muons can be used to measure the density-length of a gigantic geophysical substance, like a volcano, to learn about its inner structure.

Among various muon-science studies so far realized up to the end of the

2

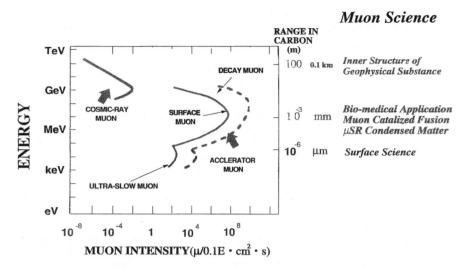

Figure 1. Two types of muons (accelerator producing and cosmic-ray) characterized by the ranges of energies, intensities, stopping-range in carbon and related scientific research fields.

20th century, we would like to focus, in this short review, on a description of the present status and expected future developments of the following three subjects, all of which have made remarkable progress recently: a) muon catalysis for nuclear fusion for a possible contribution to the next generation of nuclear energy; b) new life science studies by means of the method of labelled electrons with muons to probe electron-transfer phenomena in various biological maro-molecules; c) the construction of a data-base for disaster prevention, such as the prediction of volcanic eruptions as well as earthquakes by probing the inner structure of geological substances by using, now, cosmic-ray muons and, in the future, intense TeV muons from a possible mobile accelerator and muon-produced neutrinos.

2 Muon-Catalyzed Fusion and Fusion Energy Development

The basic process of muon-catalyzed fusion (μCF) in a D-T mixture, as depicted in the upper part of Fig.2, can be summarized as follows. After high-energy μ^- injection and stopping in a D-T mixture, either a (dμ) or a (tμ)

atom is formed, with a probability more or less proportional to the relative concentrations, D and T. Because of the difference between (dμ) and (tμ) in the binding energies of their atomic states, μ^- in (dμ) undergoes a transfer reaction to t, yielding (tμ) during a collision with the surrounding t in either D-T or T_2 molecules. The thus-formed (tμ) reacts with D_2, DT or T_2 to form a muonic molecule at a rate of $\lambda_{dt\mu}$, followed by a fusion reaction occurring from a low-lying molecular state of the (dtμ), in which the distance between d and t is sufficiently close to allow fusion to take place; a 14-MeV neutron and a 3.6-MeV α particle are emitted. After the fusion reaction inside the (dtμ) molecule, most of the μ^- are liberated, and can thus participate in a second μCF cycle. There is, however, some small fraction of the μ^- which are captured by the recoiling positively charged α. The probability of forming an $(\alpha\mu)^+$ ion is called the initial sticking ω_s^0. Once the $(\alpha\mu)^+$ is formed, the μ^- can be stripped from the $(\alpha\mu)^+$ ion, where it is "stuck" and liberated again. This process is called regeneration, with a corresponding fraction, R. Thus, μ^- in the form of either a non-stuck μ^- or one regeneration from $(\alpha\mu)^+$ can participate in a second and further μCF cycles, leading to an effective sticking parameter (ω_s) of $\omega_s=(1\text{-R})\omega_s^0$.

2.1 Present Status of μCF

In a series of recent experimental studies conducted at the RIKEN-RAL muon facility, the strongest pulsed muon facility in the world, by the RIKEN-KEK-JAERI-ETL-RAL group, systematic measurements of both 14 MeV fusion neutrons and X-rays from the sticking phenomena from the μCF in high-density (liquid/solid) and high T-concentration (10-80%) D-T mixture have been performed for the first time [1,2,3]. There, as shown in the lower part of Fig.2, the following new insight has been obtained: the initial sticking, mostly exhibited in the associated X-ray intensity from the $(\mu\alpha)^+$ ion, is rather consistent with the existing theoretical predictions, while the total sticking, mostly exhibited in a loss of the fusion neutron, is substantially smaller, suggesting the existence of an anomalous-regeneration (ionization) process of the $(\mu\alpha)^+$ during a slowing-down collision in a condensed D-T mixture. A further remarkable result has been obtained for a temperature-dependent change in the observed phenomena; by changing the temperature of solid D-T from 18 K down to 5 K, the observed anomaly becomes less prominent. The obtained result indicates that by increasing the temperature of solid D-T e.g. by applying a high pressure, one can attain a break-even condition. An experiment using a high-pressure solid D-T target is now being prepared.

In order to consider the energy-production efficiency, it is required to know how much energy is needed to produce a single muon (the muon cost). There have been several discussions on optimizing the π^- production and $\pi^- \to \mu^-$ conversion processes. Using a 1 GeV/nucleon t(d) beam to bombard Li or Be nuclei, we can obtain 0.22(0.17) π^- from a single t(d). By using a large-scale superconducting solenoid with a reflecting mirror, one can expect 75% efficiency for μ^- production from a single π^-. Since π^- production is proportional to the incident t(d) beam energy, and 1 GeV produces 0.17 μ^-, one μ^- per d(t) can be produced using an energy of 6(8) GeV. By selecting the values for π^- production in a t-t collision, the eventual cheapest cost might be about 1 π^-/4 GeV and 1 μ^-/5 GeV.

On the other hand, the energy-production capability $E^{out}_{\mu CF}$ in the μCF process is determined by $E^{out}_{\mu CF}$=17.6×Y_n(MeV) in the case of D/T μCF, which has a stringent limiting factor due to the sticking probability, ω_s ; this can be expressed as $E^{out}_{\mu CF} \leq 17.6 \times \omega_s^{-1}$ (MeV). The situation in relation to $E^{out}_{\mu CF}$ is summarized in Fig.3.

2.2 Possible Future Development of μCF

Several remarks can be made on the possibilities for a further increase in the energy-production capability of D/T μCF:

a) Since the conditions so far used for the D/T target in a μCF experiment, such as the density, temperature and C_t, as well as the energy of the (tμ) atoms $E_{t\mu}$, have not been satisfactory, there might exist more favorable conditions for higher energy production; one possibility would be a μCF experiment with a higher density D/T mixture, on the order of $\phi \cong 2\phi_0$;

b) In order to increase $\lambda_{dt\mu}$, a more favorable matching condition in terms of resonant molecular formation may exist which might be accessed by exciting the molecular levels of D_2 or DT using e.g. lasers;

c) In order to decrease ω_s, or in order to increase R, several ideas have been proposed, among which the use of high-pressure solid D-T seems to be promising, according to the recent D-T experiment, as indicated above,

d) Similarly, to increase R, the use of a D/T plasma where enhanced regeneration is expected, due to an elongated $(\alpha\mu)^+$ mean-free path and the acceleration of $(\mu\alpha)^+$ using an electric field, may be worth trying;

e) Due to collisions between the $(\alpha\mu)^+$ ions and α^{++} ions from nearby μCF reactions, exotic regeneration reactions may occur in high-density μCF in D/T mixtures with an intense μ^- beam. For this purpose it is indispensable to realize a very intense muon channel, like the "super-super muon channel" [4]. With such a system, an intensity of more than 2×10^{11} μ/s

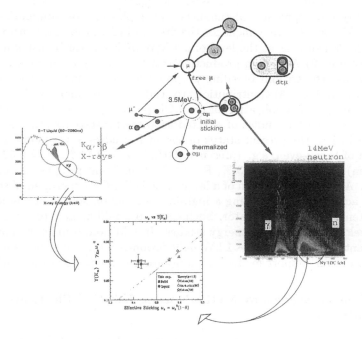

Figure 2. μCF chain reaction together with details of the fusion neutron and X-ray observables.

can be obtained with 25 MeV/c $\leq$ muon momentum $\leq$ 120 MeV/c, and one can expect an instantaneously intense muon beam of 4×10^9 μ/pulse. Using intense pulsed μ^- of this kind, muon-catalyzed fusion phenomena with an ultra-high fusion density can be hypothetically realized on a scale such as 3×10^{13} fusions/s in a ℓ target volume. The instantaneous μ^- intensity per unit volume of 2×10^8 μ^-/pulse/cc, which corresponds to a μCF density of $3 \times 10^{10} \mu$CF/pulse/cc, might be sufficient to yield interesting non-linear μCF phenomena. There, one might expect enhanced regeneration due to intense and spatially overlapping fields of the α^- particles; $(\mu\alpha)^+(I) + \alpha^{++}(II) \rightarrow \mu^- + \alpha^{++}(I) + (\alpha)^{++}(II)$. This phenomenon is similar to alpha-heating in a thermal-nuclear fusion.

A new accelerator project, such as a spallation neutron source, neutrino factory, muon colliders, etc. may significantly contribute to such progress. It is presently not difficult to realize a kW level of the μCF reactor. Considering these new trends of the μCF processes, one might expect a contribution to fusion energy development. Distinguished examples can be summarized as follows: 1) Materials development for the first wall of a fusion reactor: a high flux of 14 MeV neutrons at the level of a 1 MW μCF reactor by using a 1000 cc D-T mixture. 2) A tritium production test facility: by using high spatial density nature, the tritium breeding proposed for the fusion reactor can be more easily examined. 3) Contributions to studies of a plasma instability due to alpha-heating: the application of the μCF process just beside the plasma facility can be used for studies of selected aspects of instability studies.

In conclusion, the present status of the μCF studies is very close to the realization of break-even, e.g. (present) $150/\mu^-$, $\omega_s = 0.44$ %, R = 0.52, and (break-even) $300/\mu^-$, $\omega_s = 0.33$ %, R = 0.70. Consistent progress should be continued. Also, the realization of a large-scale D-T μCF experimental set-up should be achieved by employing a high-intensity muon source at the forthcoming intense hadron accelerator. Such a set-up would be able to contribute to the development of fusion energy studies. In addition, a quick realization of a constantly operational 0.1~1 kW μCF fusion-reactor would be called for in order to obtain support for fusion energy from individuals in the world.

3   Life Science Studies by Methods using Labelled Electrons with Muons

The electron-transfer process in macromolecules, such as proteins, is an important part of many biological phenomena, such as the storage and consummation of energy and photo-synthesis. A number of experimental investigations have been carried out to explore the electron-transfer phenomena in proteins and related chemical compounds. However, almost all of the existing information on electron transfer has been obtained by essentially macroscopic methods. In order to understand the details of electron transport, it is thus very important to use methods that can provide information at a more microscopic level.

Recently, by extending the muon spin rotation/relaxation/resonance (μSR) method we have successfully developed a method to directly observe microscopic aspects of electron transfer. The principle of μSR is based upon the particle-physics law of weak-interactions of polarized muon production via pion-decay as well as symmetric e^+/e^- emission from a polarized muon. The spin of the μ^+(positive muon), when it is born by the decay of a π^+(positive pion), is completely polarized along the direction of its motion; once the μ^+ is obtained as a beam, it is polarized along the beam direction. During slowing-down of the μ^+ inside the host material, the spin polarization is completely maintained. After thermalization and occupation at a specific microscopic location, the μ^+ decays into an e^+ and two neutrinos, and the e^+ takes a spacial distribution preferentially along the μ^+ spin. There, because the e^+ energy goes up to 50MeV, the direction of the μ^+ spin can be detected time by time by measuring the high-energy e^+ using detectors placed outside of the target material to be investigated.

In order to obtain microscopic information about electron transfer in a biological macro-molecule, the muon spin-relaxation (μSR) method offers great potential (see, Fig.4). During the slowing-down process, the injected μ^+ picks up one electron to form a neutral atomic state, called a muonium. This muonium is then thermalized, followed by chemical bonding to a reactive site on the molecule. Then, depending upon the nature of the molecule, the electron brought-in by the μ^+ can take several characteristic behaviours, including localization to form a radical state and/or a linear motion along the molecular chain. These behaviours, by setting the time-origin of electron movement, can be detected most sensitively by measuring the spin-relaxation process of the μ^+ using the μSR method, which occurs through a magnetic interaction between the μ^+ and the moving electron produced by the μ^+, itself. In other words, in place of "radioactive electrons", by introducing "electron" and an "electron observer" at the same time, the tracer of the "electron" can be made as an alternative manner: the labelled electron method.

This idea of sensitive detection of the electron behaviour in macro-molecules using muons has been successfully applied in a series of studies of electron transport in conducting polymers [5,6]. A soliton-like motion has been studied for a μ^+-produced electron in trans-polyacetylene, which contrasts with the electron localization seen in cis-polyacetylene following the formation of a radical state [5]. Similarly, polaron-type electron transport phenomena in polyaniline have also been studied [6].

In μ^+ spin-relaxation measurements under an applied longitudinal magnetic field, due to the nature of the dipole-dipole magnetic interaction between the moving electrons and the stationary muons, the characteristic dimensionality of the electron motion can be studied based on the dependence of the muon spin-relaxation rate (λ_μ) upon an externally applied magnetic field (B_{ext}); for one-dimensional electron motion, $\lambda_\mu \propto (B_{ext})^{-1/2}$, for two-dimensional electron motion, $\lambda_{\mu-} \propto (\alpha - \beta log B_{ext})$, and for three-dimensional electron motion, $\lambda_{\mu-}$ does not have a significant B_{ext} dependence [7].

Progress has been made in theoretically understanding this paramagnetic relaxation process by Risch and Kehr, who considered a stochastic treatment of the random-walk process of a spin which is rapidly diffusing along a topologically one-dimensional chain[8]. An error-function type longitudinal relaxation

8

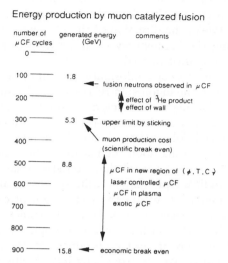

Figure 3. μCF number, energy output and remarks.

function (hereafter called the R-K function), G(t)=exp(Γt) erfc(Γt)$^{1/2}$, was proposed for $\lambda t_{max} \gg 1$, where λ is the electron spin-flip rate, t_{max} is the experimental time scale and Γ is a relaxation parameter. In this theoretical treatment, Γ is proportional to $1/B_{ext}$.

Experimental studies on a series of representative proteins conducted at the RIKEN-RAL muon facility have revealed so far the following important new aspects of electron transfer in proteins [9]:

i) For a muon-produced electron, topologically linear motion exists along the chain of both cytochrome c and myoglobin, as summarized in Fig.5. The intra-site diffusion rate along chain $D_{//}$ takes on a value on the order of 10^{12} rad/s, and is almost temperature independent. The cutoff process of the linear motion represents a departure on a longer time-scale from one-dimensional diffusion, which occurs on a shorter time scale. Thus, one can obtain the inter-chain diffusion rate, $D_{\perp}$. The obtained $D_{\perp}$ in cytochrome c is dominated by two different processes, as can be seen in the temperature dependence. The characteristic change at 200 K of $D_{\perp}$ seems to be related to the well-known structural change of a glass-like transition. On the other hand, $D_{\perp}$ in myoglobin shows only one component, reflecting the different

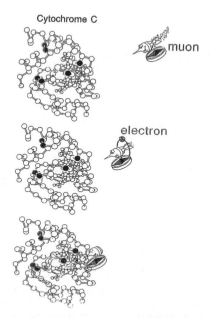

Cytochrome C

muon

electron

Figure 4. Schematic diagram representing the role of the μ^- probe for electron transfer in a macromolecule: injection of a high-energy μ^+ (top); electron pick-up and muonium (μ^+e^-) formation during slowing-down (middle); thermalization of the muonium followed by the electron release at the chemical bonding and sensitive detection of the characteristic electron motion in the macromolecule via magnetic interaction (bottom).

protein dynamics of this molecule for an inter-chain electron-transfer.

ii) Positive muons, as evidenced by preliminary paramagnetic measurements as well as theoretical calculations, stop preferentially at the negatively charged sections near to the heme-Fe atom.

iii) The characteristic electron-transfer phenomena in the form of the $1/B$ dependence was seen neither in lysozyme nor in cytochrome c with Fe(2+), where electron transfer is known not to exist.

In recent experimental studies conducted at the RIKEN-RAL muon facility, work was extended to explore electron transfer in the cytochrome c oxidase, which is known to be a terminal protein situated at the final part of the mitochondria aspiration chain, in collaboration with Drs S. Yoshikawa and K. Shinzawa-Itoh of Himeji Institute of Technology and T. Tsukihara of

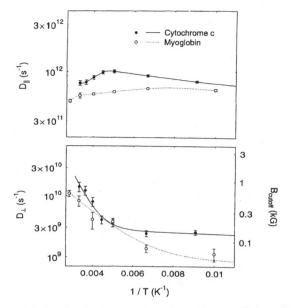

Figure 5. Upper plot showing the temperature dependence of the parallel diffusion rate of an electron in cytochrome c and myoglobin derived from the B^{-1} dependent part of the relaxation curve. The lower plot shows the perpendicular diffusion rate, derived from the cutoff field determined in Fig.6, plotted against the inverse temperature.

Institute for Protein Research, Osaka University. For this oxidase, molecular-structure studies have been carried out using high-resolution X-rays [10]. Due to a difficulty in sample preparation, the actual sample was composed of the Bovine heart cytochrome c oxidase (50%) and the surfactant material Deyl-β-D-maltoside as well as buffer chemicals of NaH_22PO_4 (altogether 50%). The results are shown in Figs. 6a and 6b in terms of the external-field dependence of the relaxation parameter (Γ) obtained by a fit to the data with the Risch-Kehr function, representing a relaxation function due to a linearly moving electron. The following conclusions were readily obtained:

1) At room temperature, electron transfer along the chain in cytochrome c oxidase is greatly suppressed compared to that in cytochrome c and myoglobin.

2) By reducing the temperature, the electron transfer along the chain becomes evident, particularly below 150K.

The results may contain contributions of the signals from either surfactant for the buffer. Since there exist several heme centers, there might be more than one corresponding signal component. Extended studies are now in progress. Electron-transfer phenomena in DNA are known to be important in view of not only damage and repair mechanisms of DNA, but also because of possible applications to new bio-devices. A recent experimental finding[11] of a possible electron transfer between G bases has accelerated both experimental and theoretical studies. Recently, at the RIKEN-RAL Muon Facility, the Yamanashi U.-RIKEN-KEK-Oxford-Tokyo Kasei Gakuen U.-Juelich collaboration, has successfully conducted a μSR experiment on oriented DNA in both A and B form DNA, and observed electron transfer in DNA, somewhat consistent with the picture of an electron hopping through base pairs [12].

4 Disasters Prevention with High-Energy Muons

For the purpose of probing the internal structure of a very large object, such as a volcanic mountain, near-horizontal cosmic-ray muons with energies higher than 100 GeV, produced as secondary cosmic rays by the interaction of the primary cosmic-ray protons in the atmospheric air, are most suitable (range of 1 TeV muon in a rock mountain: 1 km), provided that the size of the detection system is realistically large for a limited muon flux ($10^{-2}/(m^2 \cdot s \cdot str \cdot TeV)$ at 1 TeV). The basic idea of the method [13] can be summarized as follows: a) The cosmic-ray muon energy spectrum and its dependence on the vertical zenith angle are almost unique. b) The range of cosmic-ray muons through the mountain is uniquely determined by electromagnetic interactions. c) Thus, the intensity of cosmic-ray muons (N_μ^-) penetrating through rock with thickness X is uniquely determined. d) The cosmic-ray muon path through the mountain can be determined by employing a position-sensitive detection method. Based on test experiments conducted at Mt. Tsukuba and Mt. Asama, a new method for the prediction of volcanic eruptions via the detection of near-horizontal muons passing through the active part of the volcano has been proposed.

In order to overcome the intensity-limitation problem, it is indispensable to consider the use of an accelerator which is capable of approaching an active volcano. A reasonable way to obtain a mobile TeV source might be to produce muons using some modest low-energy proton accelerator, and to accelerate the thus-produced muons up to TeV by employing an acceleration scheme developed in either $\mu^+\mu^-$ colliders or in a neutrino factory. It is also interesting to point out that many of the active volcanos are situated near to the seaside. Thus, an on-ship mobile 1 TeV accelerator, if it could be realized at all, would be the most powerful tool to explore any time-dependent change in the inner structure of a volcano.

A "compact" and mobile TeV muon source can be realized by employing various new acceleration schemes and an ultra slow muon method [14]. Here, we consider the most compact scheme of a TeV muon source to be realized by the presently available technologies, and consider mounting it onto a "mobile" ship of aircraft carrier.

The scheme shown in Fig.8 is currently being proposed. For muon production, a compact proton accelerator, either a conventional synchrotron or a FFAG (fixed field alternating gradient) synchrotron up to 1 GeV would be used. There, with a minimum use of a proton linac, a compact synchrotron could be constructed within a circular space of 50 m diameter. Then, following the installation of a super-super muon channel which is a superconducting pion collector with a large acceptance and a superconducting decay solenoid [4], an ultra-slow μ^+ generation setup would be installed using hot tungsten for thermal muonium (Mu) production and laser resonant ionization of the Mu [14]. The thus-realized intense, high-quality muon ion source will have various advantages for further acceleration up to TeV energy: 1) an extremely small phase space, and 2) a small energy spread (± 0.2 eV).

A compact linac would be employed to accelerate μ^+ from 0.2 eV to 4 GeV. Then, a recirculating accelerator, as realized for the electron accelerator at CEBAF, could be used to accelerate μ^+ from 4 GeV to 1 TeV. According to a recent argument [15], the entire installation would require the space shown in Fig.7. A small energy spread is very helpful for a compact design of the "arc" for the recirculation linac. The expected performance of the proposed muon accelerator is summarized in Table 1; a 10^{10} increase could easily be realized from the value for cosmic-ray muons. The required space should be compared to the 80m × 350m available for the existing Kitty Hawk aircraft carrier.

No matter whether it is mobile or immobile, it is important that an intense TeV muon source can be realized by the scheme mentioned here. Then, by installing a relevant decay-section for the accelerated μ^+, an intense, high-quality neutrino source could be realized via $\mu^+ \to \bar{\nu}_\mu + \nu_e + e^+$, employing the scheme shown in Fig.8 [16]. Neutrinos, if generated in the form of an intense, high-quality energetic-beam, could be useful not only for particle-physics experiments, like neutrino-oscillation studies, but also for probing the inside-nature of a really gigantic substance, such as the earth, itself! Some proposals [17] exist to use neutrinos to explore the inside-nature of the earth. In this case, an intense neutrino beam is being proposed to be generated by using an intense, high-energy (multi-TeV) proton accelerator. However, because of the limited quality of the neutrino beam, the idea should be considered as unrealistic, or at least very difficult to be realized.

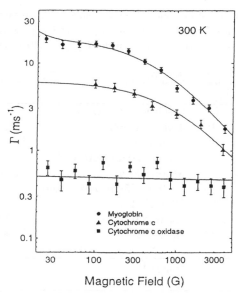

Figure 6. Observed muon spin relaxation parameter under various external magnetic fields on three typical proteins at room temperature (a). Same for cytochrome c oxidase at various temperatures (b).

The important key factors for realizing this advanced neutrino beam can be summarized as follows: i) Muon decay section. Since the decay length (L_μ) of accelerated muons is fairly long, like $L_\mu(m) \approx 4.7p_\mu$ (MeV/c), we should have an ingenious muon-decay section with a long distance and good confinement. The idea of a race-track type muon storage ring, as shown in Fig.8, is one of the most attractive ideas. ii) Quality of the produced neutrino beam. The neutrino beam produced via the decay of the accelerated muons should have properties due to the kinematics of three-body muon-decay. The energy spread as well as decay cone are the limiting factors. Regarding the decay cone, the opening half-angle, (θ_μ), becomes smaller at higher muon energy (E_{mu}), in a relation of m_μ/E_μ.

All of these properties for the case of the proposed system optimized for the future 0.8 GeV × 300μA proton synchrotron are summarized in Table 1 of the proposal [16]. With the help of a future intense pulsed proton source (with more than MW power) to also be available as a spallation neutron source, a level of $10^{12}\mu^+$/s can be obtained. As rough estimate for the other

Table 1. Expected performance of a mobile TeV muon source: $N_{\pi^+}^{tot}$, total π^+ at production target; N_{μ^+}, total μ^+ at the exit of decay solenoid; N_{μ}^{stop}, μ^+ stopping number at thermal Mu producing material; $N_{th.Mu}$, thermal Mu yield; $N_{u.s.\mu^+}$, ultra-slow μ^+ yield after thermal Mu ionization.

	$N_{expected}$ (s^{-1})	Conditions	Remarks
N_p	1.9×10^{15}	$0.8\text{GeV} \times 300\mu\text{A}$	
$N_{\pi^+}^{tot}$	2.4×10^{13}	4 cm Carbon	$\sigma_{\pi^+}^{tot}$:28mb
N_{μ^+}	6.0×10^{11}	Pion Capture	Ref.[7]
		2.9 T,20cm bore×1.5m	
		Muon Decay	p_μ: 88(42) MeV
		3.0 T, 25cm bore×10m	
N_μ^{tot}	2.9×10^{11}	$20 \times 100\mu\text{m}$ W	
N_μ^{stop}	1.2×10^{10}	$\epsilon_{th.Mu}$: 0.04	2000 K hot W
$N_{u.s.\mu^+}$	1.0×10^{10}	ϵ_{ion}: 0.8	Laser Resonant
			Ionization
$N_{\mu^+}^{Acc.}$(1TeV)	10^8	$\epsilon_{cap.}$:1.0	
		$\epsilon_{acc.}$:1.0	

proton accelerators can be obtained by scaling to the proton beam power. The arrival timing of the neutrinos and the quality of the neutrino beam can be monitored via a charged-particle appearance reaction, like $\bar{\nu}_\mu \to \mu^+$. Such a reaction cross-section becomes larger at higher energy, so that detection becomes easier for high-energy neutrinos.

There are several important applications for such an advanced neutrino beam. Some distinguished examples are neutrino oscillation, neutrino applications to the exploration of an inner-structure of the earth, and neutrino telecommunications. Among them, the importance of exploring the inside-nature of the earth is obvious. As predicted by various papers (e.g. ref.[17]), the use of neutrinos is quite promising to explore the details of the inner-structure of the earth. The neutrino intensity transmitted through the earth for a fixed neutrino energy ($\bar{E}_\nu$ depends upon the density distribution of the inner-earth structure. With the help of a variable value of the average neutrino energy ($\bar{E}_\nu \cong E_\mu/3$) by changing the muon energy ($\bar{E}_\nu$), more involved information would be obtained concerning the density as well as element distribution of the inner part of the earth. By employing the advanced neutrino beam proposed here, a time-dependent change of the inner-earth structure might be monitored, providing the most important information and a data-base for the

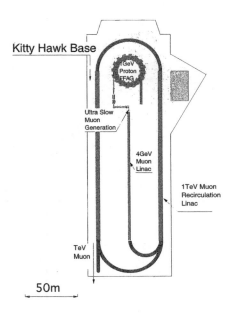

Figure 7. Mobile TeV muon source installed on the base of Kitty Hawk.

earthquake prediction.

Acknowledgements

The author wold like to express his sincere thanks to the following persons for the contributions to the latest growth of muon science: Drs T. Matsuzaki, K. Ishida, S. N. Nakamura, N. Kawamura, F. L. Pratt, I. Watanabe, E. Torikai, M. Iwasaki, K. Shimomura, Y. Miyake and Mr H. Tanaka.

References

1. K. Nagamine, T. Matsuzaki, K. Ishida et al., Hyperfine Interactions 119 (1999) 273; K. Ishida, K. Nagamine, T. Matsuzaki et al., Phys. Rev. Lett., submitted.
2. N. Kawamura, K. Nagamine, Phys. Lett. B465 (1999) 74.
3. S. N. Nakamura, K. Nagamine et al., Phys. Lett. B473 (2000) 226.

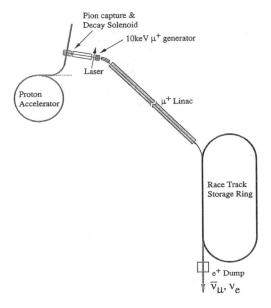

Figure 8. Proposed scheme of the advanced generation of neutrino beam based upon a decay of muons accelerated from an ultra-slow μ^+ ion-source.

4. K. Ishida and K. Nagamine, KEK Proceedings 98-5 II & JHF-98-2 II (1998) 12.
5. K. Nagamine, K. Ishida, H. Shirakawa, et al., Phys. Rev. Lett. 53, (1984) 1763; K. Ishida K. Nagamine, H. Shirakawa, et al., Phys. Rev. Lett. 55, (1985) 2209.
6. F. L. Pratt et al., Phys. Rev. Lett. 79, (1997) 2855.
7. M. A. Butler, L. R. Walker and Z. G. Soos. J. Chem. Phys. 64, (1976) 3592.
8. R. Risch and K. W. Kehr. Phys. Rev. B46, (1992) 5246.
9. K. Nagamine, F. L. Pratt et al., Physica B 289-290, (2000) 631.
10. T. Tsukihara et al., Science 269, (1995) 1069.
11. E. Meggers et al., J. Am. Chem. Soc., 120, (1998) 12950.
12. E. Torikai, K. Nagamine, H. Urabe et al., to be published (2001).

13. K. Nagamine et al., Nucl. Instr. A356 (1995) 585.
14. K. Nagamine et al., Phys. Rev. Lett. 74 (1995) 4811.
15. J. S. Berg, AIP Conference Proceedings 580 (2000) 13.
16. K. Nagamine, Proc. Japan academy 75 (1999) 255.
17. A. De Rújula, S. L. Glashow, R. R. Wilson and G. Charpak: Physics Reports 6, (1983) 341-396.

THEORETICAL MOTIVATIONS FOR
LEPTON FLAVOR VIOLATION

JONATHAN L. FENG

Center for Theoretical Physics, Massachusetts Institute of Technology
Cambridge, MA 02139 USA

In the coming years, experiments underway will increase the sensitivity to charged lepton flavor violation by four orders of magnitude. These experiments will stringently probe weak scale physics. I review the status of global symmetries in the standard model and present several well-motivated models that predict observable lepton flavor violation. Finally, I describe what we might learn from future experimental results, whether positive or null.

1 Introduction

The diverse topics discussed at this conference, ranging from charged lepton flavor violation (LFV) to neutrino oscillations to proton decay, have been brought together by a common experimental dependence on high intensity muon and neutrino sources. From a theoretical perspective, however, these topics also have a unifying theme — they are all probes of global symmetries. The exploration of symmetries and symmetry breaking is, of course, central to particle physics. These experiments form a comprehensive program to probe the standard model's (broken) global symmetries, just as high energy colliders aim to study thoroughly the standard model's (broken) local symmetries.

Global symmetries are brittle. It is therefore somewhat surprising that global symmetry breaking is currently observed only in neutrino oscillations. In the near future, however, experiments will increase the sensitivity to LFV by four or more orders of magnitude. These experiments have the potential to uncover many exotic phenomena, which may exist at a variety of scales.[1] What is even more compelling, however, at least to me, is that these experiments will for the first time stringently probe the weak scale, where we have strong reasons to expect new physics to appear. There is thus the promise of a fascinating interplay between high precision LFV experiments and high energy experiments probing electroweak symmetry breaking. For this reason, results from upcoming LFV experiments, whether positive or null, will have interesting implications.

I begin with a brief review of the standard model's global symmetries in Sec. 2. Section 3 contains a simple but telling model-independent analysis of the reach of present and future experiments. From this it will be clear that future LFV experiments will be strong probes of new physics at the weak scale. I will then focus on some broad possibilities related to the physics of electroweak symmetry breaking in Sec. 4, and consider the example of supersymmetry in Sec. 5. In Sec. 6, I conclude with several illustrations of what we might learn from future results.

2 Global Symmetries of the Standard Model

The fields of the standard model, their spins, and their quantum numbers under the three standard model gauge symmetries are:

	q	u	d	l	e	h	
Spin S	$\frac{1}{2}$	$\frac{1}{2}$	$\frac{1}{2}$	$\frac{1}{2}$	$\frac{1}{2}$	0	
SU(3)	3	3	3	1	1	1	(1)
SU(2)	2	1	1	2	1	2	
U(1)	$\frac{1}{6}$	$\frac{2}{3}$	$-\frac{1}{3}$	$-\frac{1}{2}$	-1	$\frac{1}{2}$	

Renormalizability and gauge invariance restrict the interactions among these fields to the form

$$\mathcal{L}_{\text{Yukawa}} = y_{e_{ij}} h \bar{l}_i e_j + y_{d_{ij}} h \bar{q}_i d_j + y_{u_{ij}} h^c \bar{q}_i u_j \ , \tag{2}$$

where i and j are generational indices. Through unitary rotations in generation space, we may set $y_{e_{ij}} = \text{diag}(y_e, y_\mu, y_\tau)$. The Yukawa couplings, then, conserve baryon number B, total lepton number L, and the lepton flavor numbers L_e, L_μ, and L_τ. These are the global symmetries of the standard model.

Global symmetries are broken in many possible ways, however. Beginning with the most robust theoretically, even in the standard model itself, non-perturbative quantum effects[2] break most of the global symmetries, preserving only $B - L$. Gravitational interactions are also expected to break all global symmetries through, for example, processes involving black holes. Both of these effects are typically extremely suppressed, but demonstrate that global symmetries are by no means sacred. More experimentally relevant examples arise in attempts to extend the standard model. In grand unified theories, for example, additional gauge bosons lead to global symmetry violation: in SU(5) only $B - L$ is preserved, and in SO(10), even this is broken. Finally, as we will see below, attempts to explain electroweak symmetry breaking often predict some form of global symmetry violation. In many cases, these predictions are generically at levels far beyond current constraints. It is a triumph of the standard model, and a source of frustration for those attempting to extend it, that its 'accidental' global symmetries are so accurately obeyed in nature.

3   Mass Scales and Experimental Probes

What is the current experimental status of global symmetries? For a model-independent answer, we turn to the formalism of effective field theory. If we expect the standard model to be valid up to some energy scale M, the effects of physics above this scale may be accounted for by shifts in the standard model couplings and a series of non-renormalizable terms suppressed by powers of M.[3] The terms of Eq. (2) are then to be viewed as merely the first in a series. At higher orders in $1/M$, one finds non-renormalizable operators that violate the global symmetries:

$$B-\text{violating}: \quad \frac{qqql}{M^2} \ , \ \dots \tag{3}$$

$$L-\text{violating}: \quad \frac{hhll}{M} \ , \ \dots \tag{4}$$

$$L_i-\text{violating}: \quad \frac{h\,\bar{l}\sigma^{\alpha\beta}e\,F_{\alpha\beta}}{M^2} \ , \ \frac{\bar{l}l\bar{e}e}{M^2} \ , \ \dots \ , \tag{5}$$

where generational indices have been omitted. The experimental status of global symmetries may be roughly summarized by determining what scales M are currently being probed.

The most stringent probe of baryon number violation is proton decay. Given the operator of Eq. (3), the proton lifetime is $\tau_p \propto M^4$. Current bounds on the proton lifetime with respect to various decay modes are of order $\tau_p \gtrsim 10^{32}$ years and imply $M \gtrsim 10^{15}$ GeV.

Evidence for total lepton number violation is provided by non-vanishing neutrino masses. The operator of Eq. (4) yields a neutrino mass when electroweak symmetry is broken and the Higgs scalar obtains a vacuum expectation value. For $\langle h \rangle \sim m_W$ and $m_\nu^2 \sim \Delta m_{\text{atm}}^2 \sim 10^{-3}$ eV^2, the scale of L violation seen in neutrino oscillation experiments is $M \sim 10^{14}$ GeV.

Neutrino oscillations also provide evidence for LFV at high scales, but do not preclude much larger L_i-violating, but L-conserving, effects in the charged lepton sector. For charged LFV, there are a wide variety of probes,[1,4] including $\mu \to e\gamma$, $\mu - e$ conversion, $\mu \to eee$, $\tau \to \mu\gamma$, $Z \to \mu\tau$, $K \to \mu e$, and others. All of these may be useful (see below). Here, let's concentrate on two of the most promising: $\mu \to e\gamma$, and $\mu - e$ conversion. To estimate the reach of current and future experiments, consider a refined version of the first operator of Eq. (5):

$$\mathcal{L}_{12} = e\,\frac{g^2}{16\pi^2}\frac{m_\mu}{M_{12}^2}\,\bar{\mu}\sigma^{\alpha\beta}e\,F_{\alpha\beta} \ . \tag{6}$$

The additional factors reflect an expectation that the new physics enters at one-loop and is also suppressed by a chirality insertion proportional to m_μ.

With this parametrization, the rates for $\mu \to e\gamma$ and $\mu - e$ conversion are

$$\frac{B(\mu \to e\gamma)}{1.2 \times 10^{-11}} = \left[\frac{20 \text{ TeV}}{M_{12}}\right]^4 \qquad \text{(MEGA)}$$

$$\frac{R(\mu N \to eN)}{6.1 \times 10^{-13}} = \left[\frac{10 \text{ TeV}}{M_{12}}\right]^4 \qquad \text{(SINDRUM)}$$

$$\frac{B(\mu \to e\gamma)}{1 \times 10^{-14}} = \left[\frac{110 \text{ TeV}}{M_{12}}\right]^4 \qquad (\mu e\gamma \text{ at PSI}) \qquad (7)$$

$$\frac{R(\mu N \to eN)}{5 \times 10^{-17}} = \left[\frac{110 \text{ TeV}}{M_{12}}\right]^4 \qquad \text{(MECO)}$$

$$\frac{R(\mu N \to eN)}{1 \times 10^{-18}} = \left[\frac{280 \text{ TeV}}{M_{12}}\right]^4 \qquad \text{(PRISM)} \,.$$

In the first two lines the rates are normalized to the best current bounds from the MEGA[5] and SINDRUM[6] collaborations. The last three are normalized to the expected sensitivities of the $\mu^+ \to e^+\gamma$ experiment at PSI,[7] the MECO experiment at Brookhaven,[8] and the proposed KEK/JAERI joint project, PRISM.[9] Here $R(\mu N \to eN) \equiv \Gamma(\mu N \to eN)/\Gamma(\mu N \to \text{ capture})$, and I have assumed $R(\mu N \to eN) \approx B(\mu \to e\gamma)/300$. The latter relation depends on the target nucleus N,[10] but this approximation is adequate for our purposes. More importantly, I have assumed that the new physics enters solely through the operator of Eq. (6). The sensitivity of $\mu - e$ conversion relative to $\mu \to e\gamma$ may be very much greater in other cases.

From Eq. (7), two conclusions may be drawn. First, experiments underway will extend the sensitivity in mass scale probed by roughly one order of magnitude. This is remarkable, given that the rates scale as $1/M^4$. Second, the near future sensitivity will be well above the weak scale. Many exotic particles may mediate LFV. These include extra families, Z' bosons, and extra scalars. All of these particles may exist at or above the weak scale, and so provide motivation for LFV experiments.[11]

Particularly compelling, however, is that LFV experiments will soon stringently probe the weak scale. Current theories of electroweak symmetry breaking, which are necessarily tied to the weak scale, are therefore of great relevance for LFV experiments. I will discuss some of these possibilities in the following section. Note that in this analysis, I have already included factors for loop and chirality suppression. In specific models these may be supplemented by additional suppressions from, for example, GIM-type cancellations or small intergenerational mixing angles. A more detailed analysis runs counter to the model-independent spirit of this section. In concrete examples, however, these factors typically serve only to reduce the $\mathcal{O}(100 \text{ TeV})$ reach to a more typical weak scale value of $\mathcal{O}(1 \text{ TeV})$, and observable LFV rates from weak scale physics are still predicted.

4 LFV and Electroweak Symmetry Breaking

The standard model and two possible frameworks for extending it nicely illustrate some of the possibilities for global symmetry violation:

$$
\begin{array}{ccccc}
\text{Example} & \text{Standard Model} & \text{Extra Dims} & \text{SUSY} & \\
M_{\text{High}} & B, L, L_i & & B, L & \quad (8) \\
M_{\text{Weak}} & & B, L, L_i & L_i &
\end{array}
$$

Here I have indicated the scales at which the various global symmetries are generically broken. $M_{\text{Weak}} \sim 100$ GeV is the scale of electroweak symmetry breaking. M_{High} represents some much higher scale, such as the scale of right-handed neutrinos $M_N \sim 10^{14}$ GeV, grand unification $M_{\text{GUT}} \approx 2 \times 10^{16}$ GeV, or gravitational interactions $M_{\text{Planck}} \sim 10^{19}$ GeV.

As previously noted, in the standard model, even extended to include neutrino masses, all global symmetries are broken only at some high scale. In this case, LFV experiments are far from the sensitivity required, and the exploration of global symmetry violation will be confined to neutrino and proton decay experiments. Note that the observed neutrino mixings do induce charged LFV at the loop level. However, for $\Delta m_\nu^2 \sim 10^{-3}$ eV^2, the induced LFV is $B(l \to l'\gamma) \sim 10^{-48}$, far beyond reach.

There are, however, several reasons to expect the standard model to break down well below M_{High}. Prominent among these is the gauge hierarchy problem, but there are many others, including the need for sufficient baryogenesis, and the necessity of additional non-baryonic dark matter. In addition, if current indications for a Higgs boson with mass $m_h \approx 115$ GeV are valid, the standard model vacuum will be destabilized at energy scales $\sim 10^6$ GeV, indicating that some new physics must enter below this scale.[12]

Extra dimensions have been suggested as one way to address the gauge hierarchy problem.[13] With extra dimensions, the gravitational force law is modified at distances small compared to the size of the extra dimensions. For particular choices of the number and size of the extra dimensions, the fundamental gravitational scale may be lowered to near the weak scale, translating the gauge hierarchy problem into a problem of hierarchies in length scales.

In such scenarios, one typically expects all global symmetries to be broken at $M_{\text{grav}} \sim M_{\text{Weak}}$, leading to, among other things, catastrophic proton decay.

The challenge of finding an elegant and compelling solution to this difficulty, particularly one consistent with baryogenesis, is one of the important problems in this scenario.

It is possible, however, to assume that proton stability is achieved by some mechanism and consider the possibility of LFV at low scales. This approach has been taken by a number of authors.[14] They find that in certain scenarios constraints from LFV are among the most stringent. Large rates at future experiments may therefore be expected in models with low scale gravity, especially if the gravity scale is not far above the weak scale.

In supersymmetry, a hierarchy between B and L violation on the one hand and L_i flavor violation on the other arises as an immediate consequence of the basic motivations for supersymmetry. As is well-known, two of the most important phenomenological motivations for supersymmetry are the gauge hierarchy problem and the existence of dark matter. Supersymmetry stabilizes the gauge hierarchy by introducing at the weak scale a scalar superpartner $\tilde{f}$ for every standard model fermion f (and also a fermionic superpartner for every standard model boson). These new states introduce many additional renormalizable and gauge-invariant interactions. In the notation of superfields, the terms analogous to the standard model's Yukawa terms of Eq. (2) are given by the superpotential

$$W = y_e H_d LE + y_d H_d QD + y_u H_u QU$$
$$+\lambda LLE + \lambda' LQD + \lambda'' UDD . \tag{9}$$

Here generational labels are suppressed, and each superfield contains both scalar and fermion fields: $F \supset (f, \tilde{f})$. Interactions are formed by choosing any combination of two fermions and one scalar from any term.

It is not difficult to see that these new interactions destroy the beautiful properties of the standard model with respect to global symmetries. In particular, the terms in the second line of Eq. (9) violate both B and L. However, these terms also allow all superpartners to decay, destroying the possibility of supersymmetric dark matter. This important virtue may be preserved by requiring R-parity conservation, where $R \equiv (-1)^{B+L+2S}$. This eliminates all interactions in the second line of Eq. (9), and so B and L are again broken only at a high scale. As we will see below, however, the remaining supersymmetric interactions still allow L_i-violating processes at the weak scale, leading to the hierarchy displayed in Eq. (8).

5 LFV in Supersymmetry

Because LFV experiments will probe weak scale effects, new physics that is intimately tied to the weak scale is of special importance. I now consider several examples in supersymmetry, which provides a concrete and quantitative framework for evaluating the prospects for LFV.

Supersymmetric theories are specified by the interactions of Eq. (9), and a number of supersymmetry breaking terms. Of the latter, the most important for our present purposes are slepton masses

$$m^2{}_{ij}^{LL} \tilde{l}_i^* \tilde{l}_j + m^2{}_{ij}^{RR} \tilde{e}_i^* \tilde{e}_j \, , \tag{10}$$

where the m^2 are a priori arbitrary Hermitian matrices, and trilinear terms

$$A_{e\,ij} h_d \tilde{l}_i^* \tilde{e}_j \, , \tag{11}$$

which, after electroweak symmetry breaking, couple left- and right-handed charged sleptons through the mass matrix $m^2{}_{ij}^{LR} \equiv A_{e\,ij}\langle h_d\rangle$.

These terms lead to charged LFV. To see this, it is convenient to choose the basis in which the standard model leptons are rotated to diagonalize the lepton Yukawa coupling, and the sleptons are rotated to preserve flavor-diagonal gaugino couplings. No additional flavor rotations are then available to diagonalize the terms of Eqs. (10) and (11), and the off-diagonal elements in the slepton mass matrices mediate charged LFV. Define parameters $\delta_{ij}^{MN} \equiv m^2{}_{ij}^{MN}/\tilde{m}^2$, where $\tilde{m}^2$ is the representative slepton mass scale, $M, N = L, R$, and i, j are generational indices. The branching ratios for $l \to l'\gamma$, normalized to current bounds,[5,15] constrain these parameters[a] through[16]

$$\frac{B(\mu \to e\gamma)}{1.2 \times 10^{-11}} \sim \max\left[\left(\frac{\delta_{12}^{LL,RR}}{2.0 \times 10^{-3}}\right)^2 , \left(\frac{\delta_{12}^{LR}}{6.9 \times 10^{-7}}\right)^2\right] \left[\frac{100 \text{ GeV}}{\tilde{m}}\right]^4$$

$$\frac{B(\tau \to e\gamma)}{2.7 \times 10^{-6}} \sim \max\left[\left(\frac{\delta_{13}^{LL,RR}}{2.2}\right)^2 , \left(\frac{\delta_{13}^{LR}}{1.3 \times 10^{-2}}\right)^2\right] \left[\frac{100 \text{ GeV}}{\tilde{m}}\right]^4 \tag{12}$$

$$\frac{B(\tau \to \mu\gamma)}{1.1 \times 10^{-6}} \sim \max\left[\left(\frac{\delta_{23}^{LL,RR}}{1.4}\right)^2 , \left(\frac{\delta_{23}^{LR}}{8.3 \times 10^{-3}}\right)^2\right] \left[\frac{100 \text{ GeV}}{\tilde{m}}\right]^4 \, .$$

Here I have assumed the lightest neutralino to be photino-like with mass $m_{\tilde{\gamma}}^2/\tilde{m}^2 = 0.3$; the bounds are fairly insensitive to this ratio.

---

[a] More precisely, they constrain $(\delta_{ij}^{MN})_{\text{eff}} \sim \max\left[\delta_{ij}^{MN}, \delta_{ik}^{MP}\delta_{kj}^{PN}, \ldots, (i \leftrightarrow j)\right]$.

The bounds from all three decays are non-trivial, but are especially stringent for transitions between the first and second generations. Clearly, arbitrary slepton mass matrices are forbidden. This is a statement of the (leptonic) supersymmetric flavor problem. For supersymmetry to be viable, some structure in the supersymmetry breaking masses must be present. I now turn to a variety of possibilities and analyze the LFV consequences of each.

5.1 SU(5)

The unification of gauge couplings is a strong motivation to consider supersymmetric grand unified theories. In the case of SU(5), the fields of each generation are contained in two multiplets, the $\mathbf{10}$ and $\bar{\mathbf{5}}$, while the up- and down-type Higgs doublets are contained in the $\mathbf{5}_H$ and $\bar{\mathbf{5}}_H$ multiplets, respectively. The Yukawa couplings are given by the superpotential

$$W_{\text{SU(5)}} = y_{u_{ij}}\mathbf{10}_i\mathbf{10}_j\mathbf{5}_H + y_{d_{ij}}\bar{\mathbf{5}}_i\mathbf{10}_j\bar{\mathbf{5}}_H \ . \tag{13}$$

What is the form of the scalar masses? A simple and conservative assumption is that they are $m_0^2\widetilde{\mathbf{10}}_i^*\,\widetilde{\mathbf{10}}_i + m_0^2\widetilde{\bar{\mathbf{5}}}_i^*\widetilde{\bar{\mathbf{5}}}_i$, universal, and therefore flavor-conserving, at M_{Planck}. However, to determine the physical consequences, we must evolve these to low energy scales. In the renormalization group (RG) evolution above the GUT scale, grand unified interactions will generate off-diagonal slepton masses.[17] This is easily seen by noting that y_d may be diagonalized with rotations on the $\mathbf{10}$ and $\bar{\mathbf{5}}$ fields, but there is then no remaining freedom to diagonalize y_u. RG evolution will then generate mixing of the $\mathbf{10}$ representations, and since $\tilde{e}_R, \tilde{\mu}_R, \tilde{\tau}_R \in \mathbf{10}$, non-vanishing δ_{ij}^{RR} are generated. Detailed calculations show that for some parameter choices, $\mu_L^- \to e_R^-\gamma$ is already near experimental limits, and well within reach of future experiments.[18] Note that the left-handed sleptons are not significantly mixed, and so the branching ratio for $\mu_R^- \to e_L^-\gamma$ is negligible.

5.2 Right-handed Neutrinos

A similar analysis may be applied to supersymmetric theories with right-handed neutrinos N. The leptonic superpotential is

$$W_N = y_{e_{ij}}H_d L_i E_j + y_{\nu_{ij}}H_u L_i N_j + m_{N_{ij}}N_i N_j \ . \tag{14}$$

Assume again, conservatively, that the slepton masses are universal at the Planck scale. Rotations of the L, E, and N multiplets may diagonalize y_e and m_N, but then off-diagonal entries in y_ν remain. These will generate mixing in the L representations through RG effects above the right-handed neutrino mass scale, and since $\tilde{e}_L, \tilde{\mu}_L, \tilde{\tau}_L \in L$, non-vanishing parameters δ_{ij}^{LL} are generated, which mediate $\mu_R^- \to e_L^- \gamma$. These are again within reach of future experimental sensitivities.[19] Note that, in contrast to the case of SU(5), here no δ_{ij}^{RR} mixing is generated, and so $\Gamma(\mu_L^- \to e_R^- \gamma) \approx 0$.

5.3 Flavor Symmetries

In the previous two examples, the Yukawa couplings were taken to have the hierarchical form required by experiment, and the scalar masses were assumed to have a simple universal form, presumably following from some unspecified supersymmetry breaking mechanism. An orthogonal approach is try to understand both Yukawa couplings and scalar mass textures in terms of a single flavor symmetry. In this approach, one must first find flavor symmetries that reproduce all of the known lepton masses and neutrino parameters. Then, since these symmetries determine also the superpartner mass matrices, unambiguous predictions for charged LFV follow.

The existing data may be summarized as follows: letting $\lambda \sim 0.2$ be the small flavor breaking parameter, the charged lepton masses satisfy

$$m_\tau/\langle h_d \rangle \sim \lambda^3 - 1, \quad m_\mu/m_\tau \sim \lambda^2, \quad m_e/m_\mu \sim \lambda^3 , \tag{15}$$

and the neutrino data imply

$$\sin\theta_{23} \sim 1, \quad \sin\theta_{13} \lesssim \lambda, \quad \sin\theta_{12} \sim \begin{cases} 1 & \text{MSW(LA), VO} \\ \lambda^2 & \text{MSW(SA)} \end{cases} ,$$

$$\frac{\Delta m^2_{\nu 12}}{\Delta m^2_{\nu 23}} \sim \begin{cases} \lambda^2 - \lambda^4 & \text{MSW} \\ \lambda^8 - \lambda^{12} & \text{VO} \end{cases} , \tag{16}$$

where MSW(LA,SA) and VO denote the various solar neutrino solutions.

These data cannot be reproduced by the simplest possibility, a single U(1) symmetry. In this case, the fermion mass matrices are determined by charge assignments Q as

$$m_{e\,ij} \sim \langle h_d \rangle \lambda^{Q(\bar{l}_i) + Q(e_j) + Q(h_d)} , \quad m_{\nu\,ij} \sim \frac{\langle h_u \rangle^2}{M_N} \lambda^{Q(l_i) + Q(l_j) + 2Q(h_u)} , \tag{17}$$

with similar expressions for the scalar masses. It is not difficult to show that these textures predict that highly mixed neutrinos have similar masses, contradicting the most straightforward interpretation of Eq. (16).

However, the observed data may be explained by, for example, $Z_n \times$ U(1) symmetries, two breaking parameters with opposite charges, or holomorphic zeros.[20] In these models, the above-mentioned difficulty is almost always circumvented as follows: the m_ν matrix produces hierarchical neutrino masses, and the m_e matrix generates large neutrino mixing by requiring that the gauge and mass eigenstates of the l representations are related by large rotations. However, because the lepton doublets contains charged leptons in addition to neutrinos, these rotation also generates large misalignments between the charged leptons and sleptons, producing large charged LFV rates.[21]

It is also possible for the m_ν matrix to have a very special form such that it singlehandedly generates both large mixings and large hierarchies in neutrino masses. Such examples have been constructed.[21] Typically, however, the former possibility is realized. Large angle neutrino solutions then imply large LFV rates well within reach of future $\mu \to e\gamma$ and $\mu - e$ conversion experiments.[21] Conversely, if no signal is seen in future experiments, many supersymmetric flavor models for neutrino masses will be excluded.

6 What Will We Learn?

6.1 Precision LFV

Given the dramatic improvements expected at future experiments, it is possible not only that LFV will be discovered, but also that it will be discovered with large statistics, allowing precision studies. Comparisons of the different rates of, for example, $\mu \to e\gamma$, $\mu - e$ conversion, and $\mu \to eee$ will allow us to disentangle which of the various operators in Eq. (5) are contributing.[22]

The distributions within one mode may also be informative. With polarized muons, angular distributions in $\mu \to e\gamma$, and possibly also in $\mu - e$ conversion, may be used to disentangle the two chirality components. This will be a powerful way to distinguish between different models,[23] as illustrated by the examples in Secs. 5.1 and 5.2.

6.2 Interplay with Colliders

If LFV is discovered in the near future, it is quite possible that the LFV-inducing particles will be produced directly at high energies. LFV effects from real particle production may then also be seen at colliders. In the case of supersymmetry, sleptons produced in one flavor eigenstate may oscillate to another before decaying, leading to anomalously large rates for $e\mu + X$ production, for example.[24] In the case of two generation mixing, the cross section for such processes in the simplest cases is $\sigma(e\mu) \propto \sin^2 2\theta_{12} \, (\Delta m_{12}^2)^2 / [(\Delta m_{12}^2)^2 + 4\tilde{m}^2 \Gamma^2]$, where θ_{12} and Δm_{12}^2 are the mixing angle and mass splitting between the first two slepton generations, and $\tilde{m}$ and Γ are the average slepton mass and decay width. The rate for low energy $\mu - e$ processes has a different dependence on mixing parameters, with $B(\mu \to e) \propto \sin^2 2\theta_{12} \left[\Delta m_{12}^2/\tilde{m}^2\right]^2$. Thus, the combination of high energy and high precision experiments may allow us to determine both mixing angles and mass splittings independently, which would be impossible with only one type of experiment.

6.3 Null Results

If no signal is seen, the implications depend on what is discovered at high energy colliders. If only a standard model-like Higgs boson is discovered there, the null LFV results will simply give powerful constraints on a variety of phenomena. Of course, in this case, large questions will remain concerning the physics of electroweak symmetry breaking.

More likely, some part of the physics of electroweak symmetry breaking will be uncovered at colliders. We have seen that, on very general grounds, theories of electroweak symmetry breaking often predict LFV effects within reach of the upcoming experiments. Null results from these experiments, then, will provide highly non-trivial guides to understanding these theories. In the case of supersymmetry, for example, null results may have far-reaching implications, excluding many currently attractive supergravity theories and favoring theories that generate extremely degenerate superpartners at a low scale.

Acknowledgments

I thank Y. Kuno and W. Molzon for the invitation to participate in this stimulating conference, and I am grateful to Y. Nir and Y. Shadmi for collaboration in the work described in Sec. 5.3. This work was supported in part by funds provided by the U. S. Department of Energy under cooperative research agreement DF–FC02–94ER40818.

References

1. For comprehensive reviews, see Y. Kuno and Y. Okada, hep-ph/9909265; and references therein.
2. G. 't Hooft, Phys. Rev. Lett. **37**, 8 (1976); Phys. Rev. D **14**, 3432 (1976).
3. T. Appelquist and J. Carazzone, Phys. Rev. D **11**, 2856 (1975).
4. W. Molzon, these proceedings.
5. M. L. Brooks *et al.* [MEGA Collaboration], Phys. Rev. Lett. **83**, 1521 (1999) [hep-ex/9905013].
6. P. Wintz, these proceedings.
7. J. Yashima, these proceedings.

8. J. Sculli, these proceedings; V. Tumakov, these proceedings.
9. M. Aoki, these proceedings; Y. Kuno, these proceedings.
10. See, for example, A. Czarnecki, W. J. Marciano and K. Melnikov, hep-ph/9801218; Y. G. Kim, P. Ko, J. S. Lee and K. Y. Lee, Phys. Rev. D **59**, 055018 (1999) [hep-ph/9811211]; T. S. Kosmas, A. Faessler, F. Simkovic and J. D. Vergados, Phys. Rev. C **56**, 526 (1997) [nucl-th/9704021]; A. Czarnecki, these proceedings; T. S. Kosmas, these proceedings.
11. W. Marciano, these proceedings.
12. M. Sher, Phys. Rept. **179**, 273 (1989).
13. I. Antoniadis, N. Arkani-Hamed, S. Dimopoulos and G. Dvali, Phys. Lett. **B436**, 257 (1998) [hep-ph/9804398].
14. See, for example, A. E. Faraggi and M. Pospelov, Phys. Lett. **B458**, 237 (1999) [hep-ph/9901299]; A. Ioannisian and A. Pilaftsis, Phys. Rev. D **62**, 066001 (2000) [hep-ph/9907522].
15. K. W. Edwards *et al.* [CLEO Collaboration], Phys. Rev. D **55**, 3919 (1997); S. Ahmed *et al.* [CLEO Collaboration], Phys. Rev. D **61**, 071101 (2000) [hep-ex/9910060].
16. F. Gabbiani, E. Gabrielli, A. Masiero and L. Silvestrini, Nucl. Phys. **B477**, 321 (1996) [hep-ph/9604387].
17. R. Barbieri and L. J. Hall, Phys. Lett. **B338**, 212 (1994) [hep-ph/9408406].
18. J. Hisano, T. Moroi, K. Tobe and M. Yamaguchi, Phys. Lett. **B391**, 341 (1997) [hep-ph/9605296]; J. Hisano, D. Nomura, Y. Okada, Y. Shimizu and M. Tanaka, Phys. Rev. D **58**, 116010 (1998) [hep-ph/9805367]; Y. Okada, these proceedings.
19. J. Hisano and D. Nomura, Phys. Rev. D **59**, 116005 (1999) [hep-ph/9810479]; D. Nomura, these proceedings.
20. See, for example, Y. Grossman, Y. Nir and Y. Shadmi, JHEP**9810**, 007 (1998) [hep-ph/9808355]; Y. Nir and Y. Shadmi, JHEP**9905**, 023 (1999) [hep-ph/9902293].
21. J. L. Feng, Y. Nir and Y. Shadmi, Phys. Rev. D **61**, 113005 (2000) [hep-ph/9911370].
22. A. de Gouvea, S. Lola and K. Tobe, Phys. Rev. D **63**, 035004 (2001) [hep-ph/0008085]; K. Tobe, hep-ph/0008075; K. Tobe, these proceedings.
23. Y. Kuno and Y. Okada, Phys. Rev. Lett. **77**, 434 (1996) [hep-ph/9604296].
24. N. Arkani-Hamed, H. Cheng, J. L. Feng and L. J. Hall, Phys. Rev. Lett. **77**, 1937 (1996) [hep-ph/9603431]; *ibid.*, Nucl. Phys. **B505**, 3 (1997) [hep-ph/9704205]; M. Tanaka, these proceedings.

LEPTON FLAVOR VIOLATION AND SUPERSYMMETRIC MODELS WITH RIGHT-HANDED NEUTRINO

DAISUKE NOMURA

Theory Division, KEK, Tsukuba, Ibaraki 305-0801, JAPAN
and
Dept. of Physics, Univ. of Tokyo, Tokyo 113-0033, JAPAN
E-mail: daisuke.nomura@kek.jp

In this article we review lepton flavor violation in supersymmetric models with right-handed neutrinos which naturally explain the tiny neutrino masses.

1 Introduction

Recent results of neutrino oscillation experiments provide us more and more evidences of neutrino masses and mixing[1]. Since the minimal Standard Model (SM) cannot explain the neutrino masses by itself, we are forced to extend it. The most natural and simplest extension to explain the tiny neutrino masses is the seesaw mechanism[2]. In the mechanism we introduce right-handed neutrinos which have superheavy Majorana masses among themselves and couple with left-handed neutrinos through Dirac masses which are typically of order of the weak scale. By integrating out the right-handed neutrinos we are left with Majorana mass terms of left-handed neutrino of order of v^2/M_R, where $v \simeq 246$GeV is the vacuum expectation value of the Higgs boson and M_R the right-handed Majorana mass scale. If $M_R \simeq 10^{14}$GeV we can obtain without any fine-tuning the tiny neutrino mass of order of 0.1 eV, which is consistent with the atmospheric neutrino result[1].

One of important consequences of the neutrino oscillation is the (charged-) lepton flavor violation (LFV). Description of the neutrino oscillations necessarily entails prediction of lepton flavor violating decay of charged lepton at some level. However, if we extend SM only with the right-handed neutrinos, the predicted branching ratio of $\mu \to e\gamma$ is at most 10^{-40}[3]. On the other hand, in the supersymmetric (SUSY) extensions of SM we have enough chance to expect sizable LFV rates. In SUSY SM there exist sleptons, which are new carriers of lepton flavor number. We also have to introduce mass-squared matrix of them that describe soft SUSY breaking, which is new source of LFV[4]. Since the typical scale of SUSY particle masses is the weak scale, the predicted rate is suppressed only by negative powers of SUSY particle masses. In the minimal supergravity (SUGRA) scenario in which we assume universality of the soft mass matrix at the gravitational scale $M_{\mathrm{grav}} \simeq 10^{18}$GeV, LFV off-diagonal elements can be induced radiatively via LFV interaction of physics beyond SM at high energy region[5,6,7,8,9]. Then to some extent the origin of neutrino masses can be related to the pattern of the slepton mass matrix and consequently to the LFV decay rates such as $\mathrm{Br}(\mu \to e\gamma)$ or $\mathrm{Br}(\tau \to \mu\gamma)$.

2 Lepton Flavor Violation in Supersymmetric Models with Right-handed Neutrinos

Here we first introduce the right-handed neutrinos into SUSY SM and review how the off-diagonal elements of the left-handed slepton mass matrix are radiatively generated. The superpotential of the lepton sector is given as

$$W = f_{\nu_{ij}} H_2 \overline{N}_i L_j + f_{e_{ij}} H_1 \overline{E}_i L_j + \frac{1}{2} M_{\nu_i \nu_j} \overline{N}_i \overline{N}_j. \tag{1}$$

Here H_1 and H_2 are the chiral superfields of Higgs bosons. $\overline{E}, L$ and $\overline{N}$ are chiral superfields of right-handed charged lepton, left-handed lepton, and right-handed neutrino, respectively. i and j are generation indices and run from 1 through 3. By suitable redefinition of the fields we can take the basis in which the following relations hold,

$$f_{\nu_{ij}} = f_{\nu_i} V_{Dij},$$
$$f_{e_{ij}} = f_{e_i} \delta_{ij},$$
$$M_{\nu_i \nu_j} = U_{ik}^* M_{\nu_k} U_{kj}^\dagger. \tag{2}$$

Here V_D and U are unitary matrices. The mass matrix of left-handed neutrinos is written in terms of V_D and U as

$$(m_\nu)_{ij} = V_{Dik}^\top (\overline{m}_\nu)_{kl} V_{Dlj}, \tag{3}$$

where

$$(\overline{m}_\nu)_{ij} = m_{\nu_i D} \left[M^{-1} \right]_{ij} m_{\nu_j D}$$
$$\equiv V_{Mik}^\top m_{\nu_k} V_{Mkj}. \tag{4}$$

Here, $m_{\nu_i D} = f_{\nu_i} v \sin\beta / \sqrt{2}$ and V_M is a unitary matrix. The observed mixing angles at the atmospheric[1] and the solar[10] neutrino experiments are $(V_M V_D)_{\tau\mu}$ and $(V_M V_D)_{\mu e}$, respectively, if they come from the oscillations of $\nu_\mu - \nu_\tau$ and $\nu_e - \nu_\mu$.

The SUSY breaking terms for the lepton sector in SUSY SM with the right-handed neutrinos are in general given as

$$-\mathcal{L}_{\text{SUSY breaking}} = (m_{\tilde{L}}^2)_{ij} \tilde{l}_{Li}^\dagger \tilde{l}_{Lj} + (m_{\tilde{e}}^2)_{ij} \tilde{e}_{Ri}^* \tilde{e}_{Rj} + (m_{\tilde{\nu}}^2)_{ij} \tilde{\nu}_{Ri}^* \tilde{\nu}_{Rj}$$
$$+ (A_\nu^{ij} h_2 \tilde{\nu}_{Ri}^* \tilde{l}_{Lj} + A_e^{ij} h_1 \tilde{e}_{Ri}^* \tilde{l}_{Lj} + \frac{1}{2} B_\nu^{ij} \tilde{\nu}_{Ri}^* \tilde{\nu}_{Rj}^* + h.c.), \tag{5}$$

where $\tilde{l}_L$, $\tilde{e}_R$, and $\tilde{\nu}_R$ represent the left-handed slepton, and the right-handed charged slepton, and the right-handed neutrino. Also, h_1 and h_2 are the doublet Higgs bosons. In the minimal SUGRA scenario the SUSY breaking masses for sleptons, squarks, and the Higgs bosons are universal at M_{grav}, and the SUSY breaking parameters associated with the supersymmetric Yukawa couplings or masses (A or B parameters) are proportional to the Yukawa coupling constants or masses. Then, the SUSY breaking parameters in Eq. (5) are given as

$$(m_{\tilde{L}}^2)_{ij} = (m_{\tilde{e}}^2)_{ij} = (m_{\tilde{\nu}}^2)_{ij} = \delta_{ij} m_0^2,$$
$$A_\nu^{ij} = f_{\nu_{ij}} a_0, \quad A_e^{ij} = f_{e_{ij}} a_0, \quad B_\nu^{ij} = M_{\nu_i \nu_j} b_0, \tag{6}$$

at the tree level. However, these relations are subject to radiative corrections, and at the weak scale LFV off-diagonal components of soft SUSY breaking parameters are radiatively induced as

$$(m_{\tilde{L}}^2)_{ij} \simeq -\frac{1}{8\pi^2}(3m_0^2 + a_0^2) V_{Dki}^* V_{Dlj} f_{\nu_k} f_{\nu_l} U_{km}^* U_{lm} \log \frac{M_{\text{grav}}}{M_{\nu_m}},$$
$$(m_{\tilde{e}}^2)_{ij} \simeq 0,$$
$$A_e^{ij} \simeq -\frac{3}{8\pi^2} a_0 f_{e_i} V_{Dki}^* V_{Dlj} f_{\nu_k} f_{\nu_l} U_{km}^* U_{lm} \log \frac{M_{\text{grav}}}{M_{\nu_m}}, \tag{7}$$

where $i \neq j$.

3 Atmospheric Neutrino Results and $\tau \to \mu\gamma$ Decay

In this section we discuss the branching ratio of $\tau \to \mu\gamma$ using the atmospheric neutrino result. From the zenith-angle dependence of ν_e and ν_μ fluxes measured by the Super-Kamiokande it is natural that the atmospheric neutrino anomaly comes from the neutrino oscillation between ν_μ and ν_τ, and the neutrino mass-squared difference and mixing angle are expected as

$$\Delta m_{\nu_\mu \nu_\tau}^2 \simeq 10^{-(2-3)} \text{eV}^2, \quad \sin^2 2\theta_{\nu_\mu \nu_\tau} \gtrsim 0.8. \tag{8}$$

Assuming that the neutrino masses are hierarchical as $m_{\nu_\tau} \gg m_{\nu_\mu} \gg m_{\nu_e}$, the tau neutrino mass m_{ν_τ} is given as $m_{\nu_\tau} \simeq (3 \times 10^{-2} - 1 \times 10^{-1})\text{eV}$ and if the tau neutrino Yukawa coupling constant f_{ν_τ} is as large as that of the top quark, the right-handed tau neutrino M_{ν_τ} is about 10^{14-15}GeV.

In order to evaluate the $\tau \to \mu\gamma$ rate, we need the value of $V_{D\tau\mu}$, which is not necessarily the same as the $\sin\theta_{\nu_\mu \nu_\tau}$. However, it is expected that it is also of the order of one as explained below.

Let us consider only the tau and the mu neutrino masses for simplicity. In this case we parameterize two unitary matrices V_D and V_M as

$$V_D = \begin{pmatrix} \cos\theta_D & \sin\theta_D \\ -\sin\theta_D & \cos\theta_D \end{pmatrix}, \quad V_M = \begin{pmatrix} \cos\theta_M & \sin\theta_M \\ -\sin\theta_M & \cos\theta_M \end{pmatrix}. \tag{9}$$

The observed large angle $\theta_{\nu_\mu\nu_\tau}$ is a sum of θ_D and θ_M. However, in order to derive $\theta_M \sim \pi/4$ we need a fine-tuning among the independent Yukawa coupling constants and the mass parameters. The neutrino mass matrix $(\overline{m}_\nu)$ for the second and the third generations (Eq. (3)) is given as

$$(\overline{m}_\nu) \propto \frac{1}{1 - \frac{M_{\nu_\mu\nu_\tau}^2}{M_{\nu_\mu\nu_\mu}M_{\nu_\tau\nu_\tau}}} \begin{pmatrix} \frac{m_{\nu_\mu D}^2}{M_{\nu_\mu\nu_\mu}} & -\frac{m_{\nu_\mu D}m_{\nu_\tau D}}{M_{\nu_\mu\nu_\tau}} \frac{M_{\nu\ mu\nu_\tau}^2}{M_{\nu_\mu\nu_\mu}M_{\nu_\tau\nu_\tau}} \\ -\frac{m_{\nu_\mu D}m_{\nu_\tau D}}{M_{\nu_\mu\nu_\tau}} \frac{M_{\nu\ mu\nu_\tau}^2}{M_{\nu_\mu\nu_\mu}M_{\nu_\tau\nu_\tau}} & \frac{m_{\nu_\tau D}^2}{M_{\nu_\tau\nu_\tau}} \end{pmatrix}.$$

If the following relations are valid,

$$\frac{m_{\nu_\tau D}^2}{M_{\nu_\tau\nu_\tau}} \simeq \frac{m_{\nu_\mu D}^2}{M_{\nu_\mu\nu_\mu}} \simeq \frac{m_{\nu_\mu D}m_{\nu_\tau D}}{M_{\nu_\mu\nu_\tau}}, \tag{10}$$

$m_{\nu_\tau} \gg m_{\nu_\mu}$ and $\theta_M \simeq \pi/4$ can be derived. However, these relation among independent parameters are not natural without some mechanism or symmetry. Also, if $m_{\nu_\tau D} \gg m_{\nu_\mu D}$ similarly to the quark sector, the mixing angle θ_M tends to be suppressed as

$$\tan 2\theta_M \simeq 2\left(\frac{m_{\nu_\mu D}}{m_{\nu_\tau D}}\right)\left(\frac{M_{\nu_\mu\nu_\tau}}{M_{\nu_\mu\nu_\mu}}\right). \tag{11}$$

Therefore, in the following discussion we assume that the large mixing angle between ν_τ and ν_μ comes from θ_D and that V_M is a unit matrix.

Large $V_{D\tau\mu}$ leads to non-vanishing $(m_{\tilde{L}}^2)_{\tau\mu}$ as

$$(m_{\tilde{L}}^2)_{\tau\mu} \simeq \frac{1}{16\pi^2}(3m_0^2 + a_0^2)\sin 2\theta_D f_{\nu_\tau}^2 \log\frac{M_{\text{grav}}}{M_{\nu_\tau}}, \tag{12}$$

which result in sizable $\tau \to \mu\gamma$ decay rate. In Fig. 1 we show the branching ratio of $\tau \to \mu\gamma$ as a function of the Dirac neutrino mass for tau neutrino $m_{\nu_\tau D}$ (the right-handed tau neutrino masss M_{ν_τ}). Here, $m_{\nu_\tau} = 0.07$eV, $\sin 2\theta_D = 1$. Also, we take $m_{\tilde{e}_L} = 170$GeV and the wino mass 130GeV. The other gaugino masses are determined by the GUT relation for the gaugino masses. Also, we impose the radiative breaking condition of the $SU(2)_L \times U(1)_Y$ gauge symmetries with $\tan\beta = 3, 10, 30$ and the Higgsino mass parameter positive. The branching ratio is proportional to $m_{\nu_\tau D}^4$ $(M_{\nu_\tau}^2)$. If 10^{-8} can be reached in near future experiments, we can probe $m_{\nu_\tau D} > 20(80)$GeV for $\tan\beta = 30(3)$. Then, if the Dirac tau neutrino mass is as large as the top quark mass, we may observe $\tau \to \mu\gamma$.

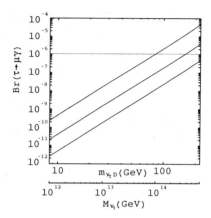

Figure 1. Dependence of the branching ratio of $\tau \to \mu\gamma$ on the Majorana mass M_{ν_τ} of the right-handed tau neutrino. Corresponding Dirac mass of the tau neutrino $m_{\nu_\tau D}$ is also shown. Here $m_{\nu_\tau} = 0.07\text{eV}$, $\sin 2\theta_D = 1$. Also, we take $m_{\tilde{e}_L} = 170\text{GeV}$ and the wino mass 130GeV. The other gaugino masses are determined by the GUT relation for the gaugino masses for simplicity. Also, we impose the radiative breaking condition of the $SU(2)_L \times U(1)_Y$ gauge symmetries with $\tan\beta = 3, 10, 30$ and the Higgsino mass parameter positive. Here also the larger $\tan\beta$ corresponds to the upper line. The horizontal dotted line is the current experimental bound.

4 Solar Neutrino Results and $\mu \to e\gamma$ Decay

In this section we discuss possible relation between the solar neutrino result and $\mu \to e\gamma$, assuming that the solar neutrino deficit comes from the $\nu_e - \nu_\mu$ oscillation.

There are three major candidates of the solution for the solar neutrino problem. The MSW solutions [11] due to the matter effect in the sun give the natural explanation, and the observation favors

$$\Delta m^2_{\nu_e \nu_Y} \simeq 10^{-(4-5)}\text{eV}^2 \quad \text{or} \quad 10^{-7}\text{eV}^2, \qquad \sin^2 2\theta_{\nu_e \nu_Y} \gtrsim 0.5, \qquad (13)$$

or

$$\Delta m^2_{\nu_e \nu_Y} \simeq 10^{-5}\text{eV}^2, \qquad \sin^2 2\theta_{\nu_e \nu_Y} \simeq 10^{-(2-3)}. \qquad (14)$$

If the solar neutrino anomaly comes from so-called 'just so' solution [12], the mass-squared difference and mixing angle are expected as [12]

$$\Delta m^2_{\nu_e \nu_Y} \simeq \times 10^{-(10-11)}\text{eV}^2, \qquad \sin^2 2\theta_{\nu_e \nu_Y} \gtrsim 0.5. \qquad (15)$$

Assuming that the neutrino masses are hierarchical as $m_{\nu_\tau} \gg m_{\nu_\mu} \gg m_{\nu_e}$, it is natural to consider ν_Y as ν_μ. If one of the large angle solutions for the solar neutrino anomaly is true, the large mixing $\theta_{\nu_\mu \nu_e}$ may imply sizable LFV mixing between sleptons of the first- and the second-generations. Similarly to the atmospheric neutrino case, it is natural to consider that the large mixing angle between ν_μ and ν_e comes from V_D.

The amplitude for $\mu \to e\gamma$ is proportional to $(m_{\tilde{L}}^2)_{\mu e}$, and it has two contributions in this model as

$$(m_{\tilde{L}}^2)_{\mu e} \simeq -\frac{1}{8\pi^2}(3m_0^2 + a_0^2) \times$$
$$\left(V_{D\tau\mu}^* V_{D\tau e} f_{\nu_\tau}^2 \log \frac{M_{\text{grav}}}{M_{\nu_\tau}} + V_{D\mu\mu}^* V_{D\mu e} f_{\nu_\mu}^2 \log \frac{M_{\text{grav}}}{M_{\nu_\mu}} \right). \quad (16)$$

Here, we assume $f_{\nu_\tau} \gg f_{\nu_\mu} \gg f_{\nu_e}$, and the term proportional to $f_{\nu_e}^2$ is neglected. Unfortunately, we have no information on $V_{D\tau e}$ and we can not evaluate the first term in Eq. (16). On the other hand, we can evaluate the second term if $V_{D\mu e}$ can be determined from the solar neutrino result. Then, in the following, we evaluate the event rate for $\mu \to e\gamma$ assuming $V_{D\tau e} = 0$. Notice that the event rate can be larger or smaller depending on the value of $V_{D\tau e}$.

Let us evaluate $\mu \to e\gamma$ rate. The forms of the amplitude and the event rate are parallel to $\tau \to \mu\gamma$. As mentioned above, if the solar neutrino anomaly comes from the MSW effect or the vacuum oscillation with the large angle, $V_{D\mu e}$ is expected to be large. This may lead to large $(m_{\tilde{L}}^2)_{\mu e}$. In Fig. 2 (a), under the condition that

$$V_D = \begin{pmatrix} \frac{1}{\sqrt{2}} & \frac{1}{2} & \frac{1}{2} \\ -\frac{1}{\sqrt{2}} & \frac{1}{2} & \frac{1}{2} \\ 0 & -\frac{1}{\sqrt{2}} & \frac{1}{\sqrt{2}} \end{pmatrix}, \quad (17)$$

we show the branching ratio of $\mu \to e\gamma$ as a function of $m_{\nu_\mu D}$ (M_{ν_μ}). We take $m_{\nu_\mu} = 4.0 \times 10^{-3}$eV, which is consistent with the MSW solution. Other input parameters are taken to be the same as in Fig. 1. The branching ratio is promotional to $m_{\nu_\mu D}^4$ ($M_{\nu_\mu}^2$). For $\tan\beta = 30(3)$, the branching ratio reaches the experimental bound[13] $\text{Br}(\mu \to e\gamma) < 1.2 \times 10^{-11}$ when $m_{\nu_\mu D} \simeq 3(10)$GeV. An experiment being prepared at PSI[14] aims to reach 10^{-14}. This corresponds to $m_{\nu_\mu D} \simeq 0.5(2)$GeV. If we take $m_{\nu_\mu} = 1.0 \times 10^{-5}$eV expected by the 'just so' solution (Fig. 2 (b)), the branching ratio becomes slightly smaller for a fixed $m_{\nu_\mu D}$ since the log factor in Eq. (7) is smaller.

If the solar neutrino anomaly comes from the MSW solution with the small mixing, we cannot distinguish whether the mixing comes from V_D or

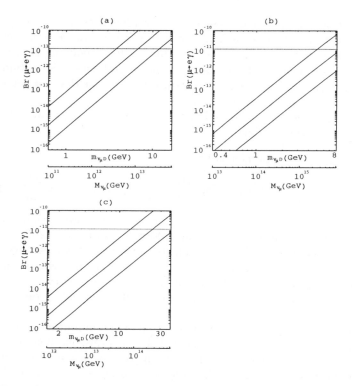

Figure 2. Dependence of the branching ratio of $\mu \to e\gamma$ on the Majorana mass M_{ν_μ} of the right-handed mu neutrino. Corresponding Dirac mass of the mu neutrino $m_{\nu_\mu D}$ is also shown. In the figures (a), (b), and (c) we take typical sample parameters for the MSW large angle solution, the vacuum oscillation solution, and the MSW small angle solution, respectively, with $V_{D\tau e} = 0$. Other parameters are the same as in the previous figure. The horizontal dotted lines are the current experimental bound.

V_M even if using argument of naturalness. If it comes from V_D, the branching ratio is smaller by about $1/100$ compared with that in the MSW large angle solution, as shown in Fig. 2 (c). In Fig. 2 (c) we assume that

$$V_D = \begin{pmatrix} 1 & 0.04 & 0.03 \\ -0.04 & 0.79 & 0.59 \\ 0 & -0.60 & 0.80 \end{pmatrix} \qquad (18)$$

and $m_{\nu_\mu} = 2.2 \times 10^{-3} \text{eV}$. Other input parameters are the same as Fig. 1.

5 Conclusions

In this article we discuss the charged lepton-flavor violating processes, $\mu \to e\gamma$ and $\tau \to \mu\gamma$, with the current neutrino experimental data in SUSY SM with right-handed neutrinos. While this model has many unknown parameters, these processes may be accessible in near future experiments. The LFV search will give new insights to the origin of the neutrino masses.

The author is supported by Research Fellowships of the Japan Society for the Promotion of Science for Young Scientists.

References

1. Y. Fukuda *et al.* (Super-Kamiokande collaboration), *Phys. Lett.* B **433**, 9 (1998); *Phys. Lett.* B **436**, 33 (1998); *Phys. Rev. Lett.* **81**, 1562 (1998).
2. T. Yanagida, in *Proceedings of the Workshop on Unified Theory and Baryon Number of the Universe*, eds. O. Sawada and A. Sugamoto (KEK, 1979) p.95;
   M. Gell-Mann, P. Ramond, and R. Slansky, in *Supergravity*, eds. P. van Nieuwenhuizen and D. Freedman (North Holland, Amsterdam, 1979).
3. T. P. Cheng and L.-F. Li, *Phys. Rev. Lett.* **45**, 1908 (1980).
4. J. Ellis and D.V. Nanopoulos, *Phys. Lett.* B **110**, 44 (1982);
   I-Hsiu Lee, *Phys. Lett.* B **138**, 121 (1984).
5. F. Borzumati and A. Masiero, *Phys. Rev. Lett.* **57**, 961 (1986).
6. J. Hisano *et al.*, *Phys. Lett.* B **357**, 579 (1995).
7. J. Hisano *et al.*, *Phys. Rev.* D **53**, 2442 (1996).
8. J. Hisano, D. Nomura, and T. Yanagida, *Phys. Lett.* B **437**, 351 (1998).
9. J. Hisano and D. Nomura, *Phys. Rev.* D **59**, 116005 (1999).
10. R. Davis, D.S. Harmer, and K.C. Hoffman, *Phys. Rev. Lett.* **20**, 1205 (1968).
11. S. P. Mikheyev and A. Y. Smirnov, *Yad. Fiz.* **42**, 1441 (1985) [*Sov. J. Nucl. Phys.* **42**, 913 (1985)]; *Nuovo Cim.* **C9**, 17 (1986);
    L. Wolfenstein, *Phys. Rev.* D **17**, 2369 (1978).
12. B. Pontecorvo, *Zh. Eksp. Teor. Fiz.* **53** 1717 (1967);
    S.M. Bilenky and B. Pontecorvo, *Phys. Rep.* **41**, 225 (1978);
    V. Barger, R.J.N. Phillips and K. Whisnant, *Phys. Rev.* D **24**, 538 (1981);
    S.L. Glashow and L.M. Krauss, *Phys. Lett.* B **190**, 199 (1987).
13. M.L. Brooks *et al.*, *Phys. Rev. Lett.* **83**, 1521 (1999).
14. L.M. Barkov *et al.*, a research proposal to PSI (1999).

LEPTON-FLAVOR VIOLATION IN SUPERSYMMETRIC MODELS WITH R-PARITY VIOLATION

KAZUHIRO TOBE

CERN, Theory Division, CH-1211 Geneva 23, Switzerland
E-mail: kazuhiro.tobe@cern.ch

Supersymmetric (SUSY) models with R-parity violation (RPV) can be an alternative scenario for non-zero neutrino masses. Within this framework, we discuss the lepton-flavor violating (LFV) processes $\mu \to e\gamma$, $\mu \to 3e$ and the $\mu \to e$ conversion in nuclei. We find the interesting features of LFV in RPV models, which are very different from other neutrino models such as SUSY models with heavy right-handed neutrinos. We show that a search for *all* the LFV processes is important to distinguish between the different models, and the measurement of P-odd asymmetries in polarized $\mu \to 3e$ is also useful to reveal the nature of LFV.

1 Introduction

Recently, Super-Kamiokande experiments on atmospheric neutrinos have announced very convincing evidence for non-zero but tiny neutrino masses. In order to accommodate such small masses, new physics beyond the standard model (SM) is necessary. The most natural scenario to account for the tiny neutrino masses is the seesaw mechanism, where the small neutrino masses are a consequence of the presence of heavy right-handed neutrinos.

Within the framework of supersymmetric (SUSY) models, there is another possible scenario to accommodate non-zero neutrino masses, that is, SUSY models with R-parity violation (RPV), in which the lepton number is broken without heavy right-handed neutrinos being introduced. Therefore it is worthwhile to consider the low-energy consequences of this framework. Here we consider, particularly, the lepton-flavor violation (LFV) in muon processes such as $\mu \to e\gamma$, $\mu \to 3e$ and the $\mu \to e$ conversion in nuclei, since the most severe constraints on certain products of RPV couplings come from the current experimental bounds on these processes[1]. In this talk, we will see the general features of LFV in SUSY models with RPV, which are quite different from those in other neutrino mass models. We will stress that, in order to distinguish between the different models, all the LFV processes are important, and P-odd asymmetries in polarized $\mu \to 3e$ are also very useful to find the nature of LFV.

In the following sections, we discuss SUSY models with RPV and neutrino masses in this framework, and then consider LFV to see some of the general features in RPV models (for all details, see Ref.[2]).

2 SUSY models with R-parity violation

First let me remind you of the supersymmetric extension of the SM. As we listed in Table 1, the doublet lepton multiplet L has the same gauge quantum numbers as the Higgs multiplet H_d, whose vacuum expectation value (vev) induces the down-type and charged lepton mass terms. If we impose only standard model gauge symmetry (without R-parity) in the model, there is no reason to distinguish a lepton doublet from a Higgs. Thus in addition to the superpotential for the ordinary Yukawa couplings, we have the following superpotential:

$$W_{RPV} = \frac{\lambda_{ijk}}{2} L_i L_j \bar{E}_k + \lambda'_{ijk} L_i Q_j \bar{D}_k + \lambda''_{ijk} \bar{U}_i \bar{D}_j \bar{D}_k + \mu'_i L_i H_u, \quad (1)$$

where the $LL\bar{E}$, $LQ\bar{D}$ and LH terms break the lepton number, and $\bar{U}\bar{D}\bar{D}$ breaks the baryon number; however, they are allowed by the SM gauge symmetry. We also have corresponding soft SUSY-breaking terms. However, the simultaneous existence of the couplings $LQ\bar{D}$ and $\bar{U}\bar{D}\bar{D}$ gives rise to rapid proton decay. The experimental limit from negative search for the proton decay provides very severe constraints on the R-parity couplings[1], for example $\lambda''_{11i}\lambda'_{11i} < 10^{-24}$ for $m_{\tilde{d}_i} = 1$ TeV. Thus we have to avoid such a rapid proton decay. One solution to the proton decay problem is "R-parity", which is the most popular solution in the SUSY SM. Doublet leptons and Higgs have different R-parity, as listed in Table 1, so that all terms in Eq. (1) are forbidden. The other solution is to impose the "baryon parity" or "lepton parity". If we require only baryon-parity conservation, the baryon-number violating term $\bar{U}\bar{D}\bar{D}$ is forbidden, and hence we can solve the proton decay problem. On the other hand, the lepton-number violating terms still exist. The existence of such a lepton-number violation will be very interesting, since it can generate non-zero neutrino masses[1].

In the next section, we briefly discuss the neutrino masses within the framework of the RPV models to see that SUSY models with RPV can be an alternative scenario for the non-zero neutrino masses.

Table 1. Gauge quantum numbers of lepton and Higgs multiplets in the minimal supersymmetric standard model

	SU(3)$_C$	SU(2)$_L$	U(1)$_Y$	R-parity
$\bar{E}$	1	1	1	-1
L	1	2	$-1/2$	-1
H_d	1	2	$-1/2$	$+1$
H_u	1	2	$1/2$	$+1$

40

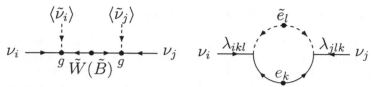

Figure 1. Feynman diagrams generating neutrino masses at tree level (left) and one-loop level (right).

3 Neutrino masses in R-parity violating models

In models with RPV, sneutrinos and neutral Higgs can in general mix, because they have the same gauge quantum numbers. Thus sneutrinos also can have vevs. As a consequence, neutrino masses can be generated at tree level via a seesaw-type mechanism mediated by a Wino ($\tilde{W}$) and Bino ($\tilde{B}$), as shown in Fig. 1. Furthermore, the trilinear RPV couplings $LL\bar{E}$ and $LQ\bar{D}$ also generate the neutrino masses at the one-loop level. In Fig. 1, we show as an example the one-loop diagram which is induced by $LL\bar{E}$ couplings. Thus, very small but non-zero sneutrino vevs and RPV couplings can explain the non-zero neutrino masses. So far many works have been done and this framework is totally consistent with the non-zero neutrino masses[1]. The interesting point is that the RPV models can generate non-zero neutrino masses *without introducing heavy right-handed neutrinos*, and hence this is very different from the ordinary seesaw mechanism. Therefore the RPV models can be an alternative scenario to accommodate non-zero neutrino masses.

Now a question arising is "Can we distinguish between two different scenarios for neutrino masses?" To address this question, we will consider in the next section, in particular, the LFV in the charged lepton sector.

4 Lepton-flavor violation in muon processes

In SUSY models with heavy right-handed neutrinos (without RPV), event rates for LFV processes can be within the reach of near-future experiments, as discussed by Nomura[3] at this workshop. In this section, we discuss LFV in the framework of the RPV models.

In SUSY models with the RPV, the LFV in muon processes is induced by diagrams such as in Figs. 2–4. Here we only consider, for simplicity, the trilinear RPV terms $LL\bar{E}$ and $LQ\bar{D}$ in Eq. (1)[a].

---

[a]Even if the bilinear term $\mu' LH_u$ and the corresponding SUSY-breaking terms were non-zero, their contributions to LFV would be negligible because of neutrino mass constraints.

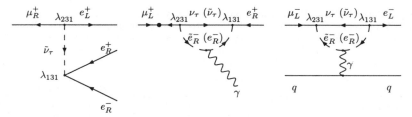

Figure 2. Feynman diagrams for LFV processes induced by $\lambda_{131}\lambda_{231}$.

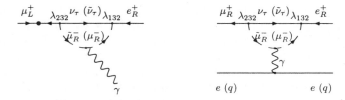

Figure 3. Feynman diagrams for LFV processes induced by $\lambda_{132}\lambda_{232}$.

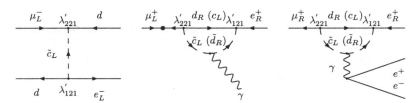

Figure 4. Feynman diagrams for LFV processes induced by $\lambda'_{121}\lambda'_{221}$.

First we show the constraints[1,2] on RPV couplings from LFV processes in Table 2, in which we assumed that only the listed pair of couplings is non-zero. The current upper limits on the branching ratios for LFV processes can put the most severe constraints on most of the listed couplings. Therefore, searches for LFV in muon processes are particularly sensitive to the RPV models. Furthermore, the future improvement of the limits on the event rates will be significant for the RPV models as shown in Table 2.

Even though results from neutrino and other experiments provide some constraints on the RPV couplings[1], it is difficult to make a definite prediction on the branching ratios for LFV processes since couplings which contribute to LFV processes are different from those which contribute to neutrino masses.

Table 2. Current (future) constraints on the R-parity violating couplings $LL\bar{E}$ and $LQ\bar{D}$ from LFV processes, assuming that only the listed pair of coupling is nonzero. The current (future) upper limits on the branching ratios are: $\mathrm{Br}(\mu \to e\gamma) < 1.2 \times 10^{-11}$ (10^{-14}), $\mathrm{Br}(\mu \to 3e) < 1.0 \times 10^{-12}$, and $\mathrm{R}(\mu \to e \text{ in Ti}) < 6.1 \times 10^{-13}$ $(\mathrm{R}(\mu \to e \text{ in Al}) < 10^{-16})$. Here we assume $m_{\tilde{\nu},\tilde{l}_R} = 100$ GeV and $m_{\tilde{q}} = 300$ GeV.

	$\mu \to e\gamma$	$\mu \to 3e$	$\mu \to e$ in nuclei
$|\lambda_{131}\lambda_{231}|$	2.3×10^{-4} (7×10^{-6})	6.7×10^{-7}	1.1×10^{-5} (2×10^{-7})
$|\lambda_{132}\lambda_{232}|$	2.3×10^{-4} (7×10^{-6})	7.1×10^{-5}	1.3×10^{-5} (2×10^{-7})
$|\lambda_{133}\lambda_{233}|$	2.3×10^{-4} (7×10^{-6})	1.2×10^{-4}	2.3×10^{-5} (4×10^{-7})
$|\lambda_{121}\lambda_{122}|$	8.2×10^{-5} (2×10^{-6})	6.7×10^{-7}	6.1×10^{-6} (1×10^{-7})
$|\lambda_{131}\lambda_{132}|$	8.2×10^{-5} (2×10^{-6})	6.7×10^{-7}	7.6×10^{-6} (1×10^{-7})
$|\lambda_{231}\lambda_{232}|$	8.2×10^{-5} (2×10^{-6})	4.5×10^{-5}	8.3×10^{-6} (1×10^{-7})
$|\lambda'_{111}\lambda'_{211}|$	6.8×10^{-4} (2×10^{-5})	1.3×10^{-4}	5.4×10^{-6} (2×10^{-7})
$|\lambda'_{112}\lambda'_{212}|$	6.8×10^{-4} (2×10^{-5})	1.4×10^{-4}	3.9×10^{-7} (7×10^{-9})
$|\lambda'_{113}\lambda'_{213}|$	6.8×10^{-4} (2×10^{-5})	1.6×10^{-4}	3.9×10^{-7} (7×10^{-9})
$|\lambda'_{121}\lambda'_{221}|$	6.8×10^{-4} (2×10^{-5})	2.0×10^{-4}	3.6×10^{-7} (6×10^{-9})
$|\lambda'_{122}\lambda'_{222}|$	6.8×10^{-4} (2×10^{-5})	2.3×10^{-4}	4.3×10^{-5} (7×10^{-7})
$|\lambda'_{123}\lambda'_{223}|$	6.9×10^{-4} (2×10^{-5})	2.9×10^{-4}	5.4×10^{-5} (9×10^{-7})

Even in this framework, however, there are some interesting features of LFV, which are not only different from those in SUSY models with heavy right-handed neutrinos, but which can also be used to characterize the different cases themselves. To identify the features, we consider three representative cases[2] in next subsections.

4.1 Case (1): $\mu^+ \to e^+e^+e^-$ is induced at tree level

First, we consider a model in which only the Yukawa couplings λ_{131} and λ_{231} are non-zero. In this case, $\mu \to 3e$ is generated at tree level, while the other LFV processes ($\mu \to e\gamma$ and $\mu \to e$ conversion in nuclei) are induced via photon penguin diagrams at the one-loop level, as shown in Fig. 2. The ratios of branching ratios, $\mathrm{Br}(\mu \to e\gamma)/\mathrm{Br}(\mu \to 3e)$ and $\mathrm{R}(\mu \to e \text{ in nuclei})/\mathrm{Br}(\mu \to 3e)$ do not depend on the RPV couplings $\lambda_{131}\lambda_{231}$, so they are more predictive quantities. For $m_{\tilde{\nu}_\tau} = m_{\tilde{e}_R} = 100$ GeV, we get

$$\frac{\mathrm{Br}(\mu \to e\gamma)}{\mathrm{Br}(\mu \to 3e)} = 1 \times 10^{-4}, \quad \frac{\mathrm{R}(\mu \to e \text{ in Ti (Al)})}{\mathrm{Br}(\mu \to 3e)} = 2\ (1) \times 10^{-3}. \quad (2)$$

Since the $\mu \to 3e$ process is generated at tree level, its branching ratio is much larger than that of the other LFV processes. If such a scenario were realized in nature, the $\mu \to 3e$ process would be a discovery mode for the LFV in muon processes. In Table 3, we list results of other similar examples.

It is very important to emphasize that the ratios of branching ratios of the different processes are very different from those in SUSY models with heavy right-handed neutrinos (with R-parity conservation), which is different neutrino mass model. In SUSY models with heavy right-handed neutrinos, the following relations are approximately satisfied:

$$\frac{\text{Br}(\mu \to e\gamma)}{\text{Br}(\mu \to 3e)} = 1.6 \times 10^2, \quad \frac{\text{R}(\mu \to e \text{ in Ti})}{\text{Br}(\mu \to 3e)} = 0.92, \tag{3}$$

since on-shell photon penguin diagram dominates over all others. Therefore, measurement of these ratios will be very important to distinguish between different neutrino mass models.

4.2   Case (2): all processes are induced at the one-loop level

Here we consider a different representative case, in which all of $\mu \to e\gamma$, $\mu \to 3e$ and $\mu \to e$ conversion in nuclei are induced at the one-loop level through the photon penguin diagram (Fig. 3). Suppose, as an example, that only the couplings λ_{132} and λ_{232} are non-zero. Then the ratios of branching ratios of the different processes are independent of the choice of $\lambda_{132}\lambda_{232}$:

$$\frac{\text{Br}(\mu \to e\gamma)}{\text{Br}(\mu \to 3e)} = 1.2, \quad \frac{\text{R}(\mu \to e \text{ in Ti (Al)})}{\text{Br}(\mu \to 3e)} = 18 \ (11), \tag{4}$$

for $m_{\tilde{\nu}_\tau} = m_{\tilde{\mu}_R} = 100$ GeV. Because of the log-enhancement of the off-shell photon penguin diagram, the event rates for $\mu \to 3e$ and $\mu \to e$ conversion in nuclei can be as large as the branching ratio for the $\mu \to e\gamma$ process, even though they are higher-order processes in QED. We also show the dependence on the slepton masses of these ratios of branching ratios in Fig. 5. All the LFV processes are equally relevant in most of the parameter space. Again we stress that these ratios of the branching ratios are very different in SUSY models with heavy right-handed neutrinos [see Eq. (3)].

4.3   Case (3): $\mu^- \to e^-$ conversion at tree level

Here, we consider the possibility that $\mu \to e$ conversion in nuclei is induced at tree level. This can arise through some of the $\lambda'LQ\bar{D}$ terms. As an example, we consider a model in which only λ'_{121} and λ'_{221} are non-zero, so that $\mu \to e$ conversion is induced at tree level, while $\mu \to e\gamma$ and $\mu \to 3e$ are generated

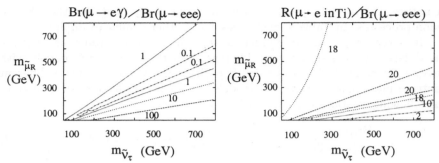

Figure 5. Contours of constant $B_{e\gamma}/B_{3e}$ (left), and $R_{conv.}/B_{3e}$ (right) in the $(m_{\tilde{\mu}_R} \times m_{\tilde{\nu}_\tau})$ plane, assuming that only the product of $LL\bar{E}$ couplings $\lambda_{132}\lambda_{232}$ is non-zero.

at one-loop level (Fig. 4). In this case, ratios of branching ratios are given by

$$\frac{\text{Br}(\mu \rightarrow e\gamma)}{\text{Br}(\mu \rightarrow 3e)} = 1.1, \quad \frac{R(\mu \rightarrow e \text{ in Ti (Al)})}{\text{Br}(\mu \rightarrow 3e)} = 2 \ (1) \times 10^5, \tag{5}$$

where we assume $m_{\tilde{d}_R} = m_{\tilde{e}_L} = 300$ GeV. Since $\mu \rightarrow e$ conversion is induced at tree level, its event rate is much larger than that of other processes, as expected. In $\mu \rightarrow 3e$, the off-shell photon penguin vertex dominates over the other contributions because of the log-enhancement. Therefore, the ratio of branching ratios $\text{Br}(\mu \rightarrow e\gamma)/\text{Br}(\mu \rightarrow 3e)$ is very similar to that we obtained in the previous subsection. Results of other similar examples are also listed in Table 3.

4.4 P-odd asymmetries in polarized $\mu^+ \rightarrow e^+e^+e^-$ process

When the muon is polarized in $\mu \rightarrow 3e$ process, two P-odd asymmetries (A_{P_1} and A_{P_2}) can be defined[4]. Following the notation introduced by Okada et al.[4], here we only show the results in Table 3. For details, see Ref.[2]. As can be seen from the table, the measurement of these P-odd asymmetries is useful to distinguish the three different representative cases in RPV models, and also important to clearly separate the different neutrino models.

5 Conclusions

We discussed the LFV of muon processes in SUSY models with RPV. Interestingly the general features of LFV are very distinct from those of SUSY models with heavy right-handed neutrinos and R-parity conservation. All the LFV processes can be very important to distinguish between different models.

Table 3. The ratios of branching ratios $\mathrm{Br}(\mu \to e\gamma)/\mathrm{Br}(\mu \to 3e)$ and $\mathrm{R}(\mu \to e$ in Ti$)/\mathrm{Br}(\mu \to 3e)$, A_{P_1} and A_{P_2} for $\mu \to 3e$ are shown when the listed pair of Yukawa couplings is dominant. Here, we assume $m_{\tilde{\nu},\tilde{l}_R} = 100$ GeV and $m_{\tilde{q}} = 300$ GeV. We also show a typical result obtained for SUSY models with heavy right-handed neutrinos and R-parity conservation.

		$\frac{B_{e\gamma}}{B_{3e}}$	$\frac{R_{conv.}}{B_{3e}}$	A_{P_1}	A_{P_2}	$\frac{A_{P_1}}{A_{P_2}}$
Case (1)	$\lambda_{131}\lambda_{231}$	1×10^{-4}	2×10^{-3}	$+19\%$	-15%	-1.3
	$\lambda_{121}\lambda_{122}$	8×10^{-4}	7×10^{-3}	-19%	$+15\%$	-1.3
	$\lambda_{131}\lambda_{132}$	8×10^{-4}	5×10^{-3}	-19%	$+15\%$	-1.3
Case (2)	$\lambda_{132}\lambda_{232}$	1.2	18	-25%	-5%	5.6
	$\lambda_{133}\lambda_{233}$	3.7	18	-25%	-4%	6.2
	$\lambda_{231}\lambda_{232}$	3.6	18	$+25\%$	$+4\%$	6.2
	$\lambda'_{122}\lambda'_{222}$	1.4	18	-25%	-4%	5.7
	$\lambda'_{123}\lambda'_{223}$	2.2	18	-25%	-4%	5.9
Case (3)	$\lambda'_{111}\lambda'_{211}$	0.4	3×10^2	-26%	-5%	5.4
	$\lambda'_{112}\lambda'_{212}$	0.5	8×10^4	-26%	-5%	5.4
	$\lambda'_{113}\lambda'_{213}$	0.7	1×10^5	-26%	-5%	5.5
	$\lambda'_{121}\lambda'_{221}$	1.1	2×10^5	-26%	-5%	5.6
MSSM with ν_R		1.6×10^2	0.92	10%	17%	0.6

Future experimental improvement of all the LFV limits will be significant to reveal the nature of LFV and the origin of neutrino masses.

Acknowledgments

The author would like to thank the organizers for a very interesting Workshop.

References

1. For a recent review on RPV, see H. Dreiner, in *Perspectives on Supersymmetry*, ed. G.L. Kane (World Scientific, Singapore, 1998). See also Ref.[2] for a complete set of references.
2. A. de Gouvêa et al., hep-ph/0008085, to appear in *Phys. Rev.* **D**.
3. See D. Nomura, contribution to this workshop.
4. Y. Okada et al., *Phys. Rev.* D **58**, 051901 (1998); **61**, 094001 (2000). See also Y. Okada, contribution to this workshop.

LFV AND FUTURE LEPTON COLLIDERS[1]

MINORU TANAKA

Department of Physics, Graduate School of Science, Osaka University,
Toyonaka, Osaka 560-0043, Japan
E-mail: tanaka@phys.sci.osaka-u.ac.jp

The observation of atmospheric neutrinos in Super-Kamiokande experiment indicates a large lepton flavor violation (LFV) among fundamental interactions in addition to neutrino masses. The minimal supersymmetric standard model augmented by right-handed neutrinos is an attractive candidate of neutrino masses and LFV. It possibly leads to a sizable mixing between the second and the third generation left-handed sleptons. In this talk, we examine probable signals of LFV in the production and decay of such left-handed sleptons at an e^+e^- linear collider and a $\mu^+\mu^-$ collider.

1 Introduction

The neutrino oscillation implied by the observation of atmospheric neutrinos by Super-Kamiokande is the first signal of lepton flavor violation (LFV) in addition to neutrino masses.[2] It suggests that fundamental interactions at a higher energy scale do not conserve the lepton flavors contrary to the standard model.

The preferred mass squared difference and mixing angle are $\Delta m^2_{\nu_\mu \nu_\tau} \simeq (5 \times 10^{-4} - 6 \times 10^{-3}) \, \text{eV}^2$ and $\sin^2 2\theta_{\nu_\mu \nu_\tau} > 0.82$. Here, we assume that ν_μ-ν_τ mixing is responsible to the atmospheric neutrino anomaly, as is consistent with the result of CHOOZ experiment.[3] Provided a mass hierarchy in the neutrino masses, we obtain that $m_{\nu_\tau} \simeq (0.02 - 0.08) \, \text{eV}$.

The simplest way to explain such a small mass of the neutrino is the seesaw mechanism. The above mass range implies that the mass scale of the right-handed neutrino is below $\sim (10^{14} - 10^{15})$ GeV as far as the neutrino Yukawa coupling is of the order of 1 or less. Combining with the large mixing angle, we expect a sizable LFV interaction between the second and the third generations below the gravitational scale ($M_G \sim 10^{18}$ GeV).

In the minimal supersymmetric standard model (MSSM), which is the most attractive extension of the standard model, the seesaw mechanism is achieved by introducing right-handed neutrino supermultiplets. Assuming the supergravity scenario of the supersymmetry (SUSY) breaking in this model, soft SUSY breaking masses of the sleptons are generated universally at the gravitational scale. The universality, however, is vitiated by the above-mentioned LFV interactions via radiative corrections below the gravitational scale. As a result the slepton mass matrices acquire LFV components. The large ν_μ–ν_τ mixing suggests large mixings between the second ($\tilde{\mu}_L$, $\tilde{\nu}_\mu$) and the third ($\tilde{\tau}_L$, $\tilde{\nu}_\tau$) generation left-handed sleptons. Slepton mixings result in several low energy LFV phenomena.[4,5,6,7,8,9] They are classified into virtual processes and real precesses. The virtual processes are loop processes in which LFV occurs in exchanging a virtual slepton. A typical example is $\tau \rightarrow \mu\gamma$. While it is possible for LFV to appear in real production and decay of sleptons.[7,8,9]

In this talk, we examine possible LFV signals in the real productions of left-handed sleptons and their subsequent decays at an e^+e^- linear collider (LC) and a $\mu^+\mu^-$ collider (MC). In Section 2, we describe the radiative generation of LFV slepton masses in the MSSM with ν_R. In Section 3, we present cross sections of several LFV signals. In Section 4, backgrounds, their reduction and sensitivity on the slepton mass difference and mixing are discussed. Section 5 is devoted to the conclusion.

2 Radiative Generation of LFV Slepton Masses in the MSSM with Right-handed Neutrinos

Adding right-handed neutrino supermultiplets to the MSSM is a simple way to explain the neutrino mass and mixing in a consistent manner with the Super-Kamiokande observation. The interpretation of the atmospheric neutrino anomaly in terms of an oscillation of ν_μ and ν_τ implies a large mixing between the second and the third generations in the lepton sector of the model. The superpotential of the lepton sector is given as

$$W_{\mathrm{MSSM}+\nu_{\mathrm{R}}} = f_{\nu_i} H_2 N_i^c L_i + f_{l_i} U_{Dij}^\dagger H_1 E_i^c L_j + \frac{1}{2} M_{ij} N_i^c N_j^c, \qquad (1)$$

where L_i's are left-handed lepton supermultiplets, and N_i^c's and E_i^c's are right-handed neutrinos and charged leptons respectively. H_1 and H_2 are the Higgs doublets in the MSSM. Here, i and j are generation indices. In the following, we concentrate on the second and third generations. The unitary matrix U_D is the leptonic counterpart of Cabibbo-Kobayashi-Maskawa matrix in the quark sector. We assume that all the couplings in Eq. (1) are real for simplicity.

We parametrize the unitary matrix as

$$U_D = \begin{pmatrix} \cos\theta_D & \sin\theta_D \\ -\sin\theta_D & \cos\theta_D \end{pmatrix}. \qquad (2)$$

The observed mixing angle between ν_μ and ν_τ is related to θ_D as $\theta_{\nu_\mu\nu_\tau} =$

48

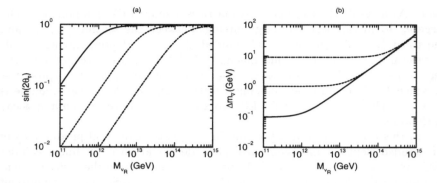

Figure 1. a) The mixing angle between the tau and muon sneutrinos ($\sin 2\theta_{\tilde\nu}$) and b) the mass difference ($\Delta m_{\tilde\nu}$) between the two non-e-like sneutrinos as functions of the right-handed neutrino scale. The input parameters are described in the text. Solid line, dashed line, and dash-dot line are for $\tan\beta = 3$, 10, and 30 respectively.

$\theta_D + \theta_M$, where θ_M is the mixing angle in the Majorana mass. In the following we assume that $\theta_M = 0$.

In the supergravity scenario, the soft SUSY breaking masses of the sleptons are flavor independent at the gravitational scale. However, because of the flavor dependence of the Yukawa couplings in Eq. (1), radiative corrections make them flavor dependent. At the weak scale, it turns out that the sneutrino mass matrix has the following form in the basis in which the charged lepton mass matrix is diagonal:

$$(m_{\tilde\nu}^2) = \begin{pmatrix} \cos\theta_{\tilde\nu} & -\sin\theta_{\tilde\nu} \\ \sin\theta_{\tilde\nu} & \cos\theta_{\tilde\nu} \end{pmatrix} \begin{pmatrix} \bar m_{\tilde\nu}^2 & 0 \\ 0 & (\bar m_{\tilde\nu} - \Delta m_{\tilde\nu})^2 \end{pmatrix} \begin{pmatrix} \cos\theta_{\tilde\nu} & \sin\theta_{\tilde\nu} \\ -\sin\theta_{\tilde\nu} & \cos\theta_{\tilde\nu} \end{pmatrix}. \quad (3)$$

The charged left-handed slepton has a similar mass matrix. In Fig. 1, we show $\sin\theta_{\tilde\nu}$ and $\Delta m_{\tilde\nu}$ as functions of the right-handed neutrino mass scale. They are obtained by numerically solving the renormalization group equations at the one-loop level.[10] Here and in the following, the minimal supergravity scenario and the GUT relation in the gaugino masses are assumed. We take $m_{\nu_\tau}^2 = 0.005$ eV^2, $\theta_D = \pi/4$, the lightest chargino mass being 100GeV, $\bar m_{\tilde\nu} = 180$GeV, $\mu > 0$ and $\tan\beta = 3$, 10, 30.

The branching ratio of $\tau \to \mu\gamma$ in the above parameter set is estimated in the range between $O(10^{-12})$ and the present experimental upper bound $[O(10^{-6})]$ depending on the right-handed neutrino mass scale and $\tan\beta$. For larger values of M_{ν_R}, which means larger $\Delta m_{\tilde\nu}$ and larger $\sin\theta_{\tilde\nu}$, B factories may find $\tau \to \mu\gamma$ process depending on the value of $\tan\beta$. For smaller M_{ν_R},

Table 1. The sample SUSY parameters in our calculation.

GeV	$m_{\tilde{\chi}_1^0}$	$m_{\tilde{\chi}_2^0}$	$m_{\tilde{\chi}_1^-}$	$m_{\tilde{\nu}_\mu}$	$m_{\tilde{\mu}_L}$	$m_{\tilde{\mu}_R}$	μ
$\tan\beta = 3$	56	105	100	180	194	159	244

however, it is unlikely. Note that the virtual processes like $\tau \to \mu\gamma$ are strongly GIM suppressed. Namely, they behaves as $\Delta m^2/m^2$. On the other hand, as is explained in the next section the real processes behave as $\Delta m^2/(m\Gamma)$. Here Γ denotes the slepton width and $\Gamma \sim 1$ GeV in the present model. Thus, it is important to search for LFV in direct production and subsequent decay of the sleptons at future colliders.

3   Cross Sections of LFV Processes in Lepton Colliders

Slepton production and decay process at a lepton collider is described by the s- and t- channel Feynman diagrams in general. An interesting point in this process is that the intermediate sleptons of different flavor interfere with each other if their mass difference is comparable to or smaller than their width. In this case, we have to incorporate the slepton width into the calculation of the cross section. Assuming the pole dominance in the phase space integration, which is legitimate when $\Delta m \lesssim \Gamma$ or $\Delta m \gg \Gamma$, the cross section may be written in the following form:

$$\sigma \simeq c^2 s^2 d[(3 - 2d)\sigma_{SS} + (3 - 2d)\sigma_{ST} + (1 - 2c^2 s^2 d)\sigma_{TT}] \qquad (4)$$

where $c = \cos\theta$ and $s = \sin\theta$ represent the slepton mixing, $\sigma_{SS(TT)}$ is the s(t)-channel cross section in the no flavor mixing case and σ_{ST} denotes the interference between s and t channels. The interference between the sleptons of different flavor is reflected in the dilution factor,

$$d = \frac{(\Delta m^2)^2}{4\bar{m}^2\Gamma^2 + (\Delta m^2)^2}. \qquad (5)$$

In the following we present some numerical results for the sample set of SUSY parameters. The input parameters are the same as Fig. 1 apart from that $\tan\beta = 3$, and $\sin\theta_{\tilde{\nu}}$ and $\Delta m_{\tilde{\nu}}$ are treated as variables. Table 1 summarizes the values of relevant parameters.

In Table 2, we show cross sections of the left-handed slepton pair production in the limit of no flavor mixing at the LC and the MC of $\sqrt{s} = 500$ GeV. Note that the LC cross sections for the second and the third generations are relatively small compared with the first generation because of the absence of

Table 2. Cross sections of the left-handed slepton production at the LC and the MC of $\sqrt{s} = 500$ GeV. We fix $m_{\tilde{\nu}}=180$ GeV, $\Delta m_{\tilde{\nu}}=1$ GeV and $\theta_{\tilde{\nu}}=0$. The other SUSY parameters are shown in Table 1.

LC	$\tilde{e}_L^+ \tilde{e}_L^-$	$\tilde{\mu}_L^+ \tilde{\mu}_L^- \ (\tilde{\tau}_L^+ \tilde{\tau}_L^-)$	$\tilde{\nu}_e \tilde{\nu}_e^*$	$\tilde{\nu}_\mu \tilde{\nu}_\mu^* \ (\tilde{\nu}_\tau \tilde{\nu}_\tau^*)$
MC	$\tilde{\mu}_L^+ \tilde{\mu}_L^-$	$\tilde{\tau}_L^+ \tilde{\tau}_L^- \ (\tilde{e}_L^+ \tilde{e}_L^-)$	$\tilde{\nu}_\mu \tilde{\nu}_\mu^*$	$\tilde{\nu}_\tau \tilde{\nu}_\tau^* \ (\tilde{\nu}_e \tilde{\nu}_e^*)$
fb	146	30	957	18

the t-channel contribution. On the other hand, the MC cross sections of the second generation slepton productions are quite large because of the t-channel contribution. In particular, the muon sneutrino cross section is almost 1 pb because of constructive interference between s and t channels.

Combining the cross sections and the branching ratios of sleptons and inos calculated with the same input parameters, we obtain the following approximate expressions of the cross sections of LFV signals at the LC:

$$\sigma(\tau^+ \mu^- + 4\,\text{jets}+ \not{E}) \simeq d(3 - 2d)\sin^2 2\theta_{\tilde{\nu}} \times 1.2\text{fb}, \tag{6}$$

$$\sigma(\tau^+ \mu^- l^\pm + 2\,\text{jets}+ \not{E}) \simeq d(3 - 2d)\sin^2 2\theta_{\tilde{\nu}} \times 0.22\text{fb}, \tag{7}$$

For the MC, we obtain

$$\sigma(\tau^+ \mu^- + 4\,\text{jets}+ \not{E}) \simeq d\sin^2 2\theta_{\tilde{\nu}}(2 - d\sin^2 2\theta_{\tilde{\nu}}) \times 30\text{fb} \tag{8}$$

$$\sigma(\tau^+ \mu^- l^\pm + 2\,\text{jets}+ \not{E}) \simeq d\sin^2 2\theta_{\tilde{\nu}}(2 - d\sin^2 2\theta_{\tilde{\nu}}) \times 5.8\text{fb} \tag{9}$$

4 Background and Sensitivity for LFV

Tau lepton pair production in the standard model processes generally provides a background for $\tilde{\tau}$-$\tilde{\mu}$ and $\tilde{\nu}_\tau$-$\tilde{\nu}_\mu$ mixings. For example, $Z^0 W^+ W^-$ production can be a background when Z decays into $\tau^+ \tau^-$ and one of the tau leptons decays into a muon. The production cross section of $Z^0 W^+ W^-$ is about 15 fb if we require central production of gauge bosons; $|\cos\theta_V| < 0.8$. Thus the background cross section is $\sigma(l^+ l^- \to Z^0(\to \tau\tau \to \mu\tau)W^+ W^-) = 0.17$fb. For our sample parameter set, $\tilde{\chi}_1^-$ and $\tilde{\chi}_2^0$ do not decay into an on-shell W or Z boson; therefore we ignore this background completely, assuming jet invariant mass cuts.

Rejecting backgrounds from the SUSY particle productions is also important for the LFV study. For the LC, the flavor conserving decays of sneutrino (left-handed charged slepton) pairs into $\tau^- \tilde{\chi}_1^+$ ($\tau^- \tilde{\chi}_2^0$) would be dangerous backgrounds. The decay of one of two τ's into μ produces $\mu\tau X$ events. 17%

Table 3. The muon misidentification probability p_μ and the tau identification probability p_τ for the IP cut of $\sigma_{IP}^{cut} = 10$ and $30\mu m$, and ∞. The sample parameter set is assumed.

σ_{IP}^{cut}	$10\mu m$	$30\mu m$	(∞)
p_μ	0.02	0.04	0.09
p_τ	0.88	0.82	0.64

of τ's decay into μ. If we cannot reject the muon, and tau leptons are identified by only the hadronic decays $[Br(\tau \to \text{hadrons}) = 64\%]$, detection of the LFV at the LC may not be very promising. For the MC, the study of LFV would be easier because of the $\tilde{\nu}_\mu \tilde{\nu}_\mu^c$ and $\tilde{\mu}^+ \tilde{\mu}^-$ productions dominate over the $\tilde{\nu}_\tau \tilde{\nu}_\tau^c$ and $\tilde{\tau}^+ \tilde{\tau}^-$ productions in the limit of small LFV.

Now we discuss cuts to eliminate $\tau\tau X$ background to $\tau\mu X$ signal. A cut on the muon energy E_μ must be very useful, and a cut on the impact parameter σ_{IP} is another possibility if a fine vertex detector is available.

The energy distribution for the signal μ is flat between $E_\mu^{\min}$ and $E_\mu^{\max}$ and the end points are the same as $E_\tau^{\min}$ and $E_\tau^{\max}$. Therefore we can reduce the $\tau \to \mu$ background by requiring $E_\mu > E_\mu^{\min}$ without costing signal events. $E_\mu^{\min}$ is 26.5 GeV in our sample parameter set.

The primary muon from the LFV signal must have a track which points back toward the interaction point. By requiring signal event to have $\sigma_{IP} < \sigma_{IP}^{cut}$ one can eliminate some backgrounds. The σ_{IP} cut is also useful to increase the efficiency to identify a tau lepton in its leptonic decay. In Table 3, we show the misidentification probability of muon and the tau identification probability when the energy cut and a σ_{IP} cut are applied. Note that the recent NLC detector study shows that $\delta\sigma = 2.4 \oplus 13.7/p\beta/\sin^{3/2}\theta(\mu m)$ is possible.[11]

Although the $\tau\mu l + 2$ jets mode is cleaner, their cross sections are significantly smaller than the $\tau\mu + 4$ jets mode. So, we discuss the sensitivity of the $\tau\mu + 4$ jets mode to LFV in the following. Fig. 2 shows the contours of constant significance corresponding to 3σ discovery in the $\sin 2\theta_{\tilde{\nu}} - \Delta m_{\tilde{\nu}}$ plane. We assume the integrated luminosity of $\mathcal{L} = 50\text{fb}^{-1}$, the proposed one-year luminosity of LCs. The muon energy cut $E_\mu > E_\mu^{\min}$ is used for both the LC and the MC cases. While we apply the impact parameter cut of $\sigma_{IP}^{cut} = 10\mu m$ only to the LC case, because the cross section is larger in the MC case and it is not known if a fine vertex detector can be placed near the interaction point of the MC.

For the LC case, the experimental sensitivity is limited by the small statistics of the signal. The parameter region of $\sin 2\theta_{\tilde{\nu}} > 0.5$ and $\Delta m_{\tilde{\nu}} > 0.4$ GeV

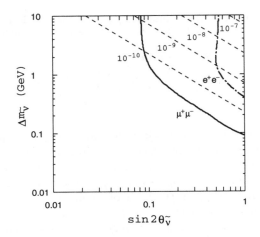

Figure 2. Significance contours corresponding to 3σ discovery in the $\sin 2\theta_{\tilde{\nu}}$–$\Delta m_{\tilde{\nu}}$ plane. The dashed-dot (solid) line is for the LC (MC) with the center mass energy 500GeV. We assume the integrated luminosity of $\mathcal{L} = 50\text{fb}^{-1}$. For the LC we take $\sigma_{IP}^{\text{cut}} = 10\mu\text{m}$. We also show the contours of constant $\tau \to \mu\gamma$ branching ratios by the dashed lines.

may be explored. For the MC case, the signal rate is much larger than the LC case. Notice that S/N in the small mixing region is enhanced due to the enhanced $\mu^+\mu^- + 4\,\text{jets}+ \not{E}$ rate over the background $\tau^+\tau^- + 4\,\text{jets}+ \not{E}$ rate. The region where $\sin 2\theta_{\tilde{\nu}} > 0.1$ and $\Delta m_{\tilde{\nu}} > 0.1$ GeV may be explored.

5   Conclusion

In this talk we discuss the detection of the LFV expected in the MSSM with right-handed neutrinos at future lepton colliders. We assume the ν_τ-ν_μ mixing inspired by the Super-Kamiokande atmospheric neutrino observation.

For the e^+e^- linear collider, the signal cross section is small because only s-channel exchange of gauge boson is involved. The background events coming from flavor conserving SUSY processes are reduced by requiring $E_\mu \geq E_\mu^{\text{min}}$. In addition, requiring σ_{IP} cut, we can improve the S/N ratio further. For the integrated luminosity of $\mathcal{L} = 50\text{fb}^{-1}$, $\sin 2\theta_{\tilde{\nu}} \gtrsim 0.5$ and $\Delta m_{\tilde{\nu}} \gtrsim 0.4\text{GeV}$ may be explored in the $\tau\mu + 4\,\text{jets}$ mode. The experimental reach is limited by statistics only. For the $\mu^+\mu^-$ collider, the signal cross sections are enhanced by the t-channel exchange of charginos or neutralinos. In the limit of the small LFV, the background from the flavor conserving production of the third generation slepton pairs is relatively suppressed since the s-channel contribution is dominant. $\sin 2\theta_{\tilde{\nu}} \gtrsim 0.1$ and $\Delta m_{\tilde{\nu}} \lesssim 0.1 \text{GeV}$ may be explored in the $\tau\mu + 4$ jets mode without the σ_{IP} cut.

In conclusion, we emphasize the advantage of the $\mu^+\mu^-$ collider in the study of LFV.

Acknowledgments

The author would like to thank his collaborators J. Hisano, M. M. Nojiri and Y. Shimizu. He also thanks the organizers of the workshop.

References

1. Based on the work by J. Hisano, M.M. Nojiri, Y. Shimizu and M. Tanaka, *Phys. Rev.* D **60**, 055008 (1999).
2. Super-Kamiokande Collaboration, Y. Fukuda *et al.*, *Phys. Rev. Lett.* **81**, 1562 (1998).
3. CHOOZ Collaboration, M. Apollonio *et al.*, *Phys. Lett.* B **420**, 397 (1998).
4. L. Hall, V. Kostelecky, and S. Raby, *Nucl. Phys.* B **267**, 415 (1986).
5. R. Barbieri and L. Hall, *Phys. Lett.* B **338**, 212 (1994); J. Hisano, T. Moroi, K. Tobe, and M. Yamaguchi, *Phys. Lett.* B **391**, 341 (1997); *Erratum-ibid* **B3973571997**; J. Hisano, D. Nomura, Y. Okada, Y. Shimizu, and M. Tanaka, *Phys. Rev.* D **58**, 116010 (1998); J. Hisano, D. Nomura, and T. Yanagida, *Phys. Lett.* B **437**, 351 (1998); Y. Okada, K. Okumura, and Y. Shimizu, *Phys. Lett.* B **58**, 051901 (1998).
6. R. Barbieri, L. Hall, and A. Strumia, *Nucl. Phys.* B **445**, 219 (1995); P. Ciafaloni, A. Romanino, and A. Strumia, *Nucl. Phys.* B **458**, 3 (1996); N. Arkani-Hamed, H.-C. Cheng, and L.J. Hall, *Phys. Rev.* D **53**, 413 (1996).
7. N. Arkani-Hamed, H.-C. Cheng, J.L. Feng, and L.J. Hall, *Phys. Rev. Lett.* **77**, 1937 (1996), *Nucl. Phys.* B **505**, 7 (1997).
8. N.V. Krasnikov, MPLA **9**, 791 (1994).
9. M. Hirouchi, and M. Tanaka, *Phys. Rev.* D **58**, 032004 (1998).
10. See for details: J. Hisano, T. Moroi, K. Tobe, and M. Yamaguchi, *Phys. Rev.* D **53**, 2442 (1996).
11. The NLC Accelerator Design Group and The NLC Physics Working Group, *Physics and Technology of the Next Linear Collider* (SLAC-Report-485).

NEUTRINO OSCILLATION SCENARIOS AND GUT MODEL PREDICTIONS

CARL H. ALBRIGHT

Fermi National Accelerator Laboratory, P.O. Box 500, Batavia, IL 60510
E-mail: albright@fnal.gov

The present experimental situation regarding neutrino oscillations is first summarized, followed by an overview of selected grand unified models which have been proposed to explain the various scenarios with three active neutrinos and their right-handed counterparts. Special attention is given to the general features of the models and their ability to favor some scenarios over others.

1 Three-Active-Neutrino Oscillation Scenarios

1.1 Atmospheric Neutrinos

Recent results from the Super-Kamiokande Collaboration [1] involving atmospheric neutrinos convincingly favor muon-neutrinos oscillating into tau-neutrinos rather than into light sterile neutrinos. The latter possibility is ruled out at the 99% confidence level. In terms of the oscillation parameters, $\Delta m_{ij}^2 \equiv m_i^2 - m_j^2$ and $\sin^2 2\theta_{atm}$, the best fit values obtained are

$$\Delta m_{32}^2 = 3.2 \times 10^{-3} \text{ eV}^2,$$
$$\sin^2 2\theta_{23} = 1.000, \tag{1}$$

with the latter related to the neutrino mixing matrix elements by $\sin^2 2\theta_{atm} = 4|U_{\mu 3}|^2|U_{\tau 3}|^2$.

1.2 Solar Neutrinos

The situation regarding solar neutrinos is considerably less certain. The recent analysis [2] by the Super-Kamiokande Collaboration involving their 1117 day sample together with the data from the Chlorine [3] and Gallium [4] experiments favor the large mixing angle MSW [5] solution (LMA) and possibly the LOW solution over the small mixing angle (SMA) and vacuum (VAC) solutions, with the latter two being ruled out at the 95% confidence level. Several theory groups analyzing the same data suggest instead that while the LMA solution is favored, the other solutions are still viable at the 95% c.l. In fact, a continuum solution – the quasi-vacuum solution (QVO) – stretches between the LOW and VAC regions with $\tan^2 \theta_{sol} \gtrsim 1.0$. The best fit points in the various parameter regions found in a recent analysis by Gonzalez-Garcia and Peña-Garay [6] are given by

$$
\begin{aligned}
SMA: \ & \Delta m_{21}^2 = 5.0 \times 10^{-6} \text{ eV}^2, \\
& \sin^2 2\theta_{12} = 0.0024, \\
& \tan^2 \theta_{12} = 0.0006, \\
LMA: \ & \Delta m_{21}^2 = 3.2 \times 10^{-5} \text{ eV}^2, \\
& \sin^2 2\theta_{12} = 0.75, \\
& \tan^2 \theta_{12} = 0.33, \\
LOW: \ & \Delta m_{21}^2 = 1.0 \times 10^{-7} \text{ eV}^2, \\
& \sin^2 2\theta_{12} = 0.96, \\
& \tan^2 \theta_{12} = 0.67, \\
QVO: \ & \Delta m_{21}^2 = 8.6 \times 10^{-10} \text{ eV}^2, \\
& \sin^2 2\theta_{12} = 0.96, \\
& \tan^2 \theta_{12} = 1.5,
\end{aligned}
\tag{2}
$$

Note that θ_{12} is in the second octant or "dark side" for the quasi-vacuum region. An even more recent analysis, [7] which includes the CHOOZ reactor constraint, [8] modifies the above numbers slightly, and sets $\tan^2 \theta_{13} = 0.005$.

1.3 Maximal and Bimaximal Mixings

The Maki-Nakagawa-Sakata (MNS) neutrino mixing matrix, analogous to the CKM mixing matrix, can be written as

$$
U_{MNS} = \begin{pmatrix} c_{12}c_{13} & s_{12}c_{13} & s_{13}e^{-i\delta} \\ -s_{12}c_{23} - c_{12}s_{23}s_{13}e^{i\delta} & c_{12}c_{23} - s_{12}s_{23}s_{13}e^{i\delta} & s_{23}c_{13} \\ s_{12}s_{23} - c_{12}c_{23}s_{13}e^{i\delta} & -c_{12}s_{23} - s_{12}c_{23}s_{13}e^{i\delta} & c_{23}c_{13} \end{pmatrix}
\tag{3}
$$

in terms of $c_{12} = \cos\theta_{12}$, $s_{12} = \sin\theta_{12}$, etc. With the oscillation parameters relevant to the scenarios indicated above, we can approximate $\theta_{13} = 0$ and $\theta_{23} = 45^o$ whereby Eq. (3) becomes essentially

$$
U_{MNS} = \begin{pmatrix} c_{12} & s_{12} & 0 \\ -s_{12}/\sqrt{2} & c_{12}/\sqrt{2} & 1/\sqrt{2} \\ s_{12}/\sqrt{2} & -c_{12}/\sqrt{2} & 1/\sqrt{2} \end{pmatrix},
\tag{4}
$$

where the light neutrino mass eigenstates are given in terms of the flavor states by

$$\nu_3 = \frac{1}{\sqrt{2}}(\nu_\mu + \nu_\tau),$$
$$\nu_2 = \nu_e \sin\theta_{12} + \frac{1}{\sqrt{2}}(\nu_\mu - \nu_\tau)\cos\theta_{12}, \tag{5}$$
$$\nu_1 = \nu_e \cos\theta_{12} - \frac{1}{\sqrt{2}}(\nu_\mu - \nu_\tau)\sin\theta_{12},$$

For the SMA solution, $\theta_{12} = 1.4^o$, while the three large mixing solar solutions differ from maximal in that the angle is approximately 30° for the LMA, 39° for the LOW, and 51° for the QVO solutions. In contrast, the CKM quark mixing matrix is approximately

$$V_{CKM} = \begin{pmatrix} 0.975 & 0.220 & 0.0032e^{-i\delta} \\ -0.220 & 0.974 & 0.040 \\ 0.0088 & -0.040 & 0.999 \end{pmatrix}. \tag{6}$$

An important issue to be answered is why $U_{\mu 3} \simeq 1/\sqrt{2}$ is so much larger than $V_{cb} \simeq 0.040$.

Maximal mixing of two neutrino mass eigenstates can arise if the two states are nearly degenerate in mass, i.e., the neutrinos are pseudo-Dirac. It can also arise if the determinant of the 2×2 submatrix nearly vanishes. For example,

$$\begin{pmatrix} x^2 & x \\ x & 1 \end{pmatrix} \rightarrow \lambda = 0, \ 1 + x^2, \ \psi_0 \sim \begin{pmatrix} 1 \\ -x \end{pmatrix}, \ \psi_+ \sim \begin{pmatrix} x \\ 1 \end{pmatrix}, \tag{7}$$

and the components are comparable for $x \simeq 1$. The first situation is relevant for the QVO and LOW near maximal mixings, while the second is more relevant for the atmospheric and LMA mixings where a sizable hierarchy is expected to be present.

Finally, it should be noted that the U_{MNS} mixing matrix is the product of two unitary transformations diagonalizing the charged lepton mass matrix L and the light neutrino mass matrix M_ν:

$$U_{MNS} = U_L^\dagger U_\nu, \tag{8}$$

where by the seesaw mechanism $M_\nu = -N^T M_R^{-1} N$ is given in terms of the Dirac neutrino matrix N and the right-handed Majorana matrix M_R. The structure of U_{MNS} is then determined by the three matrices N, M_R and L, one of which or in concert can play a role in the maximal or bimaximal mixing pattern.

2 Types of Neutrino Models and Possible Unifications

Neutrino models can be characterized as belonging to one of three types for the purpose of this talk.

- Those involving only left-handed fields leading to a left-handed Majorana mass matrix with no Dirac neutrino mass matrix present. The Zee model [9] is a prime example. Typically lepton number is violated by two units, or an $L = -2$ isovector Higgs field is introduced. A newly-defined lepton number $\bar{L} \equiv L_e - L_\mu - L_\tau$ is often taken to be conserved. The ultralight neutrino masses, however, are not easily understood.

- Models in which both left-handed and right-handed fields are present. With no Higgs contributions to the left-handed Majorana mass matrix, the seesaw mechanism readily yields ultralight neutrino masses, provided the right-handed Majorana masses are in the range of $10^5 - 10^{14}$ GeV. Such masses are naturally obtained in SUSY GUT models with $\Lambda_G = 2 \times 10^{16}$ GeV.

- Models in which neutrinos probe higher dimensions. Right-handed neutrinos which are singlets under all gauge symmetries can enter the bulk. With large extra dimensions and the compactification scale much lower than the string scale, a modified seesaw mechanism can generate ultralight neutrino masses.

I shall restrict my attention in this overview to models involving both left-handed and right-handed neutrinos. In this workshop, Tobe [10] addresses purely left-handed neutrino models in the context of R-parity violation, while Mohapatra [11] considers models involving higher dimensions.

Both the nonsupersymmetric standard model (SM) and the minimum supersymmetric extended version (MSSM) involve no right-handed neutrinos and just one or two Higgs doublets, respectively. Hence no renormalizable mass terms can be constructed for the neutrinos; moreover, the renormalizable mass terms which are present for the quarks and charged leptons have completely arbitrary Yukawa couplings. In order to reduce the number of free parameters and thereby achieve some detailed predictions for the mass spectra of the fundamental particles, some flavor and/or family unification must be introduced. This is generally done in the context of supersymmetry for which the desirable feature of gauge coupling unification obtains.

Flavor or vertical symmetry has generally been achieved in the framework of Grand Unified Theories (GUTs) which provide unified treatments of quarks and leptons, as (some) quarks and leptons are placed in the same multiplets. Examples involve $SU(5)$, $SU(5) \times U(1)$, $SO(10)$, E_6, etc.

58

The introduction of a family or horizontal symmetry, on the other hand, enables one to build in an apparent hierarchy for masses of comparable flavors belonging to different families. Such a symmetry may be discrete as in the case of Z_2, S_3, $Z_2 \times Z_2$, etc. which results in multiplicative quantum numbers. A continuous symmetry such as $U(1)$, $U(2)$, $SU(3)$, etc., on the other hand, results in additive quantum numbers and may be global or local (and possibly anomalous).

Combined flavor and family symmetries will typically reduce the number of model parameters even more effectively. On the other hand, the unification of flavor and family symmetries into one single group such as $SO(18)$ or $SU(8)$, for example, has generally not been successful, as too many extra states are present which must be made superheavy.

3 Froggatt-Nielsen-type Models with Anomalous $U(1)$ Family Symmetry

In 1979 Froggatt and Nielsen [12] added to the SM a scalar singlet "flavon" ϕ_f, which gets a VEV, together with heavy fermions, $(F, \bar{F})$, in vector-like representations, all of which carry $U(1)$ family charges. With $U(1)$ broken at a scale M_G by $\langle \phi \rangle / M_G \equiv \lambda \sim (0.01 - 0.2)$, the light and heavy fermions are mixed; hence λ can serve as an expansion parameter for the quark and lepton mass matrix entries. No GUT is involved, although M_G is some high unspecified scale.

This idea received a revival in the past decade when it was observed by Ibanez [13] that string theories with anomalous $U(1)$'s generate Fayet-Iliopoulos D-terms which trigger the breaking of the $U(1)$ at a scale of $O(\lambda)$ below the cutoff, again providing a suitable expansion parameter. The λ^n structure of the mass matrices can be determined from the corresponding Wolfenstein λ structure of the CKM matrix and the quark and lepton mass ratios, where different $U(1)$ charges are assigned to each quark and lepton field.

By careful assignment of the $U(1)$ charges, Ramond and many other authors [a] have shown that maximal mixing of $\nu_\mu \leftrightarrow \nu_\tau$ can be obtained, while the SMA solution for $\nu_e \leftrightarrow \nu_\mu$, ν_τ is strongly favored. Since then, other authors [15] have applied the technique in the presence of $SU(5)$ or $SO(10)$ to get also the QVO or LOW solutions. Very recently, Kitano and Mimura [16] have considered $SU(5)$ and $SO(10)$ models in this framework with an $SU(3) \times U(1)$ horizontal symmetry to show that the LMA solution can also be obtained. But with these types of models, the coefficients (prefactors) of the λ powers can

---

[a] For reference listings in two more comprehensive reviews, see [14].

not be accurately predicted.

4 Predictive SUSY GUT Models

With the minimal $SU(5)$ SUSY GUT model extended to include the left-handed conjugate neutrinos, the matter fields are placed in $\bar{5}$ and 10 representations according to

$$\bar{5}_i \supset (d^c_\alpha, \ell, \nu_\ell)_i, \quad 10_i \supset (u_\alpha, d_\alpha, u^c_\alpha, \ell^c)_i, \quad 1_i \supset (\nu^c)_i, \quad \alpha = 1,2,3. \quad (9)$$

while the Higgs fields are placed in the adjoint and fundamental representations

$$\Sigma(24), \quad H_u(5), \quad H_d(\bar{5}). \quad (10)$$

The $SU(5)$ symmetry is broken down to the MSSM at a scale Λ_G with $\langle\Sigma\rangle$ pointing in the $B-L$ direction, but doublet-triplet splitting must be done by hand. The electroweak breaking occurs when the H_u and H_d VEV's are generated.

The number of Yukawa couplings has now been reduced in the Yukawa superpotential, and the fermion mass matrices exhibit the symmetries, $M_U = M_U^T$, $M_D = M_L^T$. This implies $m_b = m_\tau$ at the GUT scale, but also $m_d/m_s = m_e/m_\mu$ which is too simplistic since no family symmetry is present. One can circumvent this problem by introducing a family or horizontal symmetry, but more predictive results are obtained in the $SO(10)$ framework.

In $SO(10)$ all fermions of one family are placed in a 16 spinor supermultiplet and carry the same family charge assignment:

$$16_i(u_\alpha, d_\alpha, u^c_\alpha, d^c_\alpha, \ell, \ell^c, \nu_\ell, \nu^c_\ell)_i, \quad i = 1,2,3. \quad (11)$$

Massive pairs of $(16, \overline{16})$'s and 10's may also be present. The Higgs Fields may contain one or more 45_H's and pairs of 16_H, $\overline{16}_H$ which break $SO(10)$ down to the SM, while $10_{\mathbf{H}}$ breaks the electroweak group at the electroweak scale. A $\overline{126}_H$ or effective $\overline{16}_H \cdot \overline{16}_H$ field can generate superheavy right-handed Majorana neutrino masses.

With an appropriate family symmetry introduced, a number of texture zeros will appear in the mass matrices. These will enable one to make some well-defined predictions for the masses and mixings of the quarks and leptons, for typically fewer mass matrix parameters will be present than the 20 quark and lepton mass and mixing observables plus 3 right-handed Majorana masses.

Yukawa $t - b - \tau$ coupling unification is possible only for $\tan\beta = v_u/v_d \simeq 55$ in this minimal Higgs case described above. However, if a $16'_H$, $\overline{16}'_H$ pair is introduced with the former getting an electroweak-breaking VEV which helps contribute to H_d [17], or if the $\mathbf{16}_H$ of the first pair also gets an EW VEV, Yukawa coupling unification is possible for $\tan\beta \ll 55$. Such breaking VEV's can contribute asymmetrically to the down quark and charged lepton mass matrices. This makes it possible to understand large $\nu_\mu - \nu_\tau$ mixing, $U_{\mu3} \simeq 0.707$, while $V_{cb} \simeq 0.040$. Moreover, the Georgi-Jarlskog mass relations [18],

$$m_s/m_b = m_\mu/3m_\tau \text{ and } m_d/m_b = 3m_e/m_\tau, \tag{12}$$

can be generated by the mass matrices with the help of the same asymmetrical contributions.

Models based on $SO(10)$ then differ due to their matter and Higgs contents as well as the horizontal family symmetry group chosen. Several selected illustrative examples of predictive $SO(10)$ GUT models are presented below, where some of their characteristic features are highlighted.

4.1 $SO(10)$ with $U(1)_H$

A model of this type has been presented by Babu, Pati, and Wilczek [19] based on dimension-5 effective operators involving

Matter Fields : $\mathbf{16}_1$, $\mathbf{16}_2$, $\mathbf{16}_3$
Higgs Fields : $\mathbf{10}_H$, $\mathbf{16}_H$, $\overline{\mathbf{16}}_H$, $\mathbf{45}_H$ $\qquad$ (13)

The $\mathbf{16}_H$ develops both GUT and EW scale VEV's. With no CP violation, 11 matrix input parameters yield $18 + 3$ masses and mixings. Maximal $\nu_\mu \leftrightarrow \nu_\tau$ mixing arises from the seesaw mechanism, while the SMA solar solution is preferred.

4.2 $SO(10)$ with $[U(1) \times Z_2 \times Z_2]_H$

Barr and Raby [20] have shown that a stable solution to the doublet-triplet splitting problem in $SO(10)$ can be obtained based on this global horizontal group. With an extension of the minimal Higgs content involved, Albright and Barr [21] have developed a model involving only renormalizable terms in the Yukawa superpotential with the following superfields present:

Matter Fields : $\mathbf{16}_1$, $\mathbf{16}_2$, $\mathbf{16}_3$, $2(\mathbf{16}, \overline{\mathbf{16}})'s$, $2(\mathbf{10})'s$, $6(\mathbf{1})'s$
Higgs Fields : $4(\mathbf{10}_H)'s$, $2(\mathbf{16}_H, \overline{\mathbf{16}}_H)'s$, $\mathbf{45}_H$, $5(\mathbf{1}_H)'s$ $\qquad$ (14)

Ten matrix input parameters yield all $20 + 3$ masses and mixings. A value of $\tan\beta \simeq 5$ is favored with $\sin 2\beta \sim 0.65$ obtained for the CKM unitarity triangle. Maximal $\nu_\mu \leftrightarrow \nu_\tau$ mixing arises from the lopsided texture of the charged lepton matrix. In this simplest scenario, the QVO solution is preferred, although by modifying the right-handed Majorana matrix the SMA or LMA solar solutions can be obtained with one or four more input parameters, respectively.

4.3 SO(10) with $[SU(2) \times Z_2 \times Z_2 \times Z_2]_H$

Chen and Mahanthappa [22] have based a model on this family group with dim-5 effective operators involving the following superfields:

$$
\begin{aligned}
&\text{Matter Fields} : (\mathbf{16}, 2), \ (\mathbf{16}, 1) \\
&\text{Higgs Fields} : \ 5(\mathbf{10}, 1)_H's, \ 3(\overline{\mathbf{126}}, 1)_H's \\
&\text{Flavon Fields} : 3(\mathbf{1}, 2)_H's, \ 3(\mathbf{1}, 3)_H's
\end{aligned} \tag{15}
$$

With no CP violation, eleven matrix input parameters yield $18 + 3$ masses and mixings, while $\tan\beta = 10$ is assumed. Maximal $\nu_\mu \leftrightarrow \nu_\tau$ mixing arises from the seesaw mechanism with symmetric mass matrices. The QVO solar solution is preferred, while the SMA solution or the LMA solution with $\tan^2\theta_{sol} < 1$ is difficult to obtain.

4.4 SO(10) with $[U(2) \times U(1)^n]_H$

Blazek, Raby, and Tobe [23] have constructed such a model involving only renormalizable terms in the Yukawa superpotential with the following fields:

$$
\begin{aligned}
&\text{Matter Fields} : (\mathbf{16}, 2), \ (\mathbf{16}, 1), \ (\mathbf{1}, 2), \ (\mathbf{1}, 1) \\
&\text{HiggsFields} : \ \ (\mathbf{10}, 1)_H, \ (\mathbf{45}, 1)_H \\
&\text{Flavon Fields} : 2(\mathbf{1}, 2)_H's, \ (\mathbf{1}, 3)_H, \ 2(\mathbf{1}, 1)_H's
\end{aligned} \tag{16}
$$

Sixteen matrix input parameters yield the $20 + 3$ masses and mixings with $\tan\beta \simeq 55$ required. CP violation occurs with $\sin 2\beta$ in the second quadrant for the CKM unitarity triangle. All solar neutrino solutions, SMA, LMA, LOW and QVO, are possible.

4.5 $SO(10)$ with $[SU(3) \times$ unspecified discrete symmetries$]_H$

Berezhiani and Rossi [24] have proposed a model based on this group with the following structure:

$$
\begin{aligned}
\text{Matter Fields}: & \ (\mathbf{16},\mathbf{3}),\ (\mathbf{16},\overline{\mathbf{3}}),\ (\overline{\mathbf{16}},\mathbf{3}),\ 2(\mathbf{16},\mathbf{3})'s, \\
& \ 2(\overline{\mathbf{16}},\overline{\mathbf{3}})'s,\ (\mathbf{1},\overline{\mathbf{3}}),\ (\mathbf{1},\mathbf{3}),\ (\mathbf{10},\mathbf{3}),\ (\mathbf{10},\overline{\mathbf{3}}) \\
\text{Higgs Fields}: & \ (\mathbf{16},\mathbf{1})_H,\ (\overline{\mathbf{16}},\mathbf{1})_H,\ (\mathbf{54},\mathbf{1})_H,\ 2(\mathbf{45},\mathbf{1})_H's, \\
& \ 2(\mathbf{10},\mathbf{1})_H's \\
\text{Flavon Fields}: & \ (\mathbf{1},\overline{\mathbf{6}})_H,\ 3(\mathbf{1},\mathbf{3})_H's,\ (\mathbf{1},\mathbf{8})_H
\end{aligned}
\tag{17}
$$

Fourteen matrix input parameters yield 18 + 3 masses and mixings with moderate $\tan\beta$ assumed. Maximal $\nu_\mu \leftrightarrow \nu_\tau$ mixing arises from the lopsided texture of the charged lepton mass matrix. The SMA solar solution is preferred, while other solutions are possible with modification of the right-handed Majorana matrix.

5 Concluding Remarks

The most predictive models for the 12 "light" fermion masses and their 8 CKM and MNS mixing angles and phases are obtained in the framework of grand unified models with family symmetries. The $SO(10)$ models are more tightly constrained than $SU(5)$ models and are more economical than larger groups such as E_6 [25], where more fields must be made supermassive. In fact, some $SO(10)$ models do very well in predicting the 20 + 3 "observables" with just 10 or more input parameters, depending on the model and type of solar neutrino mixing solution involved.

The SMA MSW solution is readily obtained in many unified models, since only one pair of states, ν_μ and ν_τ, are maximally mixed. Bimaximal mixing can be obtained in a smaller class of models, with the QVO solution having the more natural hierarchy with a pair of pseudo-Dirac neutrinos. The LMA MSW solution, if allowed, requires the most fine tuning, for two nearly maximal mixings must be obtained with a hierarchy of neutrino masses. For this latter solution, the right-handed Majorana neutrino masses typically span a range of $10^6 - 10^{14}$ GeV, while the lightest is typically 10^{10} GeV for the other solar neutrino solutions.

Unfortunately, while the experimental solar neutrino solution remains rather uncertain, unified model builders are not able to clarify the situation by predicting the outcome with any degree of certainty.

References

1. S. Fukuda *et al*, hep-ex/0009001.
2. T. Takeuchi, talk at *XXX Intl. Conf. on HEP*, ICHEP 2000, Osaka, Japan, July 27 - August 2, 2000.
3. K. Lande, talk at *XIX Intl. Conf. on Neutrino Physics and Astrophysics*, Sudbury, Canada, June 16-21, 2000.
4. V. Gavrin, talk at *XIX Intl. Conf. on Neutrino Physics and Astrophysics*, Sudbury, Canada, June 16-21, 2000; GALLEX Collab. (W. Hampel *et al*), *Phys. Lett.* B **447**, 127 (1999).
5. L. Wolfenstein, *Phys. Rev.* D **17**, 2369 (1978); S.P. Mikheyev and A. Yu. Smirnov, Yad. Fiz. **42**, 1441 (1985), [Sov. J. Nucl. Phys. **42**, 913 (1985)].
6. M.C. Gonzalez-Garcia and C. Peña-Garay, hep-ph/0009041.
7. M.C. Gonzalez-Garcia, M. Maltoni, C. Peña-Garay, and J.W.F. Valle, hep-ph/0009350.
8. CHOOZ Collab. (M. Apollonio *et al*), *Phys. Lett.* B **420**, 397 (1998).
9. A. Zee, *Phys. Lett.* B **93**, 389 (1980).
10. K. Tobe, contribution to this workshop.
11. R.N. Mohapatra, contribution to this workshop.
12. C.D. Froggatt and H.B. Nielsen, *Nucl. Phys.* B **147**, 277 (1979).
13. L. Ibanez, *Phys. Lett.* B **303**, 55 (1993).
14. G. Altarelli, Nucl. Phys. Proc. Suppl. **87**, 291 (2000); S.M. Barr and I. Dorsner, *Nucl. Phys.* B **585**, 79 (2000).
15. Q. Shafi and Z. Tavartkiladze, *Phys. Lett.* B **482**, 145 (2000); G. Altarelli, F. Feruglio and I. Masina, hep-ph/0007254.
16. R. Kitano and Y. Mimura, hep-ph/0008269.
17. C.H. Albright, K.S. Babu, and S.M. Barr, *Phys. Rev. Lett.* **81**, 1167 (1998).
18. H. Georgi and C. Jarlskog, *Phys. Lett.* B **86**, 297 (1979).
19. K.S. Babu, J.C. Pati, and F. Wilczek, NPB **566**, 33 (2000).
20. S.M. Barr and S. Raby, *Phys. Rev. Lett.* **79**, 4748 (1997).
21. C.H. Albright and S.M. Barr, *Phys. Rev. Lett.* **85**, 244 (2000); *Phys. Rev.* D **62**, 093008 (2000).
22. M.-C. Chen and K.T. Mahanthappa, hep-ph/0005292.
23. T. Blazek, S. Raby, and K. Tobe, *Phys. Rev.* D **62**, 055001 (2000).
24. Z. Berezhiani and A. Rossi, hep-ph/0003084.
25. E. Ma, contribution to this workshop.

The exotic $\mu^- \to e^-$ conversion in nuclei: An interplay between nuclear, particle and non-standard physics

T.S. KOSMAS

Theoretical Physics Section, University of Ioannina, GR-45110 Ioannina, Greece

The transition strengths of the $\mu^- \to e^-$ conversion channels, calculated with several nuclear methods, are exploited to put constraints on the parameters of a model dependent Lagrangian which involves interactions mediated by various particles (W-, Z-bosons, virtual photons, etc.) as well as SUSY R-parity violating interactions. We focus on the particularly interesting nuclei ^{48}Ti and ^{27}Al, i.e. the stopping targets of the $\mu^- \to e^-$ experiments SINDRUM2 and MECO.

1 Introduction

In this work we shall concentrate on the neutrinoless conversion of a negative muon, captured in the lowest atomic orbit around a nucleus, into an energetic electron,

$$\mu_b^- + (A, Z) \to e^- + (A, Z)^* , \qquad (1)$$

This process violates the muon- (L_μ) and electron- (L_e) quantum numbers and can occur by a coherent mechanism, when the bound muon changes into an electron of same total energy with the Coulomb field absorbing the excess momentum, or it can occur incoherently on a single proton in the nucleus.[1-4] This reaction offers one of the most sensitive tests of lepton flavor conservation.[5-9]

In the standard model of electroweak interactions muonic atoms decay either by muon decay in orbit, $\mu_b^- \to \nu_\mu + \bar{\nu}_e + e^-$, or by nuclear muon capture which is largely dominated by the channel

$$\mu_b^- + (A, Z) \to \nu_\mu + (A, Z - 1)^* . \qquad (2)$$

The capture probability increases as Z does and reaches a value of $\sim$0.95 for very heavy nuclei. It was found that the exotic $\mu^- \to e^-$ conversion would leave the nucleus in its ground state with probability $\sim$90%. This (coherent) process is the only measured channel without background which in the $\mu^- \to e^-$ process comes mostly from bound muon decay.[5-8]

Modern gauge theories and supersymmetric models which go beyond the standard model allow for μ_b^-, in addition to process (1), the anomalous muon-to-positron conversion, $(A, Z) + \mu_b^- \to e^+ + (A, Z-2)^*$, which violates not only L_μ and L_e but also the total-lepton (L) quantum number. The exotic channel (1) is a very good example of coexistence of nuclear and particle physics with the physics beyond the standard model. In recent years there has been intense experimental [5-9] and theoretical [10-13] effort to search for the $\mu^- \to e^-$ process, since it offers more powerful limits for the charged-lepton flavor violating (LFV) parameters compared to other purely leptonic exotic processes like $\mu \to e\gamma$, $\mu \to 3e$, etc. [2,13]

At present, there are two working $\mu^- \to e^-$ conversion experiments: (i) The PSI experiment (SINDRUM2) using ^{48}Ti as target has provided the best upper bound on the branching ratio $R_{\mu e} = \Gamma(\mu^- \to e^-)/\Gamma(\mu^- \to \nu_\mu)$, where $\Gamma(\mu^- \to \nu_\mu)$ denotes the total rate of the allowed ordinary muon-capture channel (2). The value of the best upper limit is [5]

$$R_{\mu e}^{Ti} \le 6.1 \times 10^{-13} \tag{3}$$

(ii) The designed at Brookhaven experiment (MECO) is going to employ pulsing muon beam and the light target ^{27}Al to reach a sensitivity of [7,8]

$$R_{\mu e}^{Al} \le 2 \times 10^{-17} \tag{4}$$

On the theoretical side, recently non-SUSY and R_pSUSY mechanisms have been examined, [10-13] by developing a formalism of calculating the $\mu - e$ conversion rates for quark level Lagrangians including all possible interaction terms (vector, scalar, axial-vector etc.). Special attention was devoted to the effect of nucleon and nuclear structure dependence of the branching ratio $R_{\mu e}$. The aim of the nuclear calculations is to provide the required theoretical physics aspects of the $\mu^- \to e^-$ reaction and offer a general theoretical background for the running (SINDRUM2) and planned (MECO) $\mu - e$ conversion experiments (these days the intention for designing a new $\mu - e$ conversion experiment, PRIME, at Tokyo was announced [9]). The particle physics aspects of $\mu^- \to e^-$ are involved in its effective Lagrangian written in terms of a set of lepton flavor violating parameters within new-physics extensions of the standard model. [13] With the aid of the transition matrix elements [10,11] one can constrain the LFV parameter space by making use of the existing experimental limits. [5-8]

2 Description of the $\mu^- \to e^-$ transition operators

On the underlying particle physics side there are many mechanisms of the $\mu - e$ conversion [1,4] which fall into two categories, photonic and non-photonic. These mechanisms involve different details of the nucleon and nuclear structure due to the fact that they proceed at different distances. The long-distance photonic mechanisms are mediated by virtual-photon exchange between the quark and the $\mu - e$-lepton currents. The short-distance non-photonic mechanisms are mediated by heavy particles like W, Z-bosons, [1,2] Higgs particles and super-symmetric particles (squarks, sleptons, gauginos, higgsinos etc.) [13] with and without R-parity conservation in the vertices.

The nuclear physics dependence of the $\mu^- \to e^-$ is included in the corresponding hadronic currents of the process. For some common extensions of the standard model as well as supersymmetric theories these currents involve polar-vector and axial-vector interactions parametrized as

$$J^\mu = \bar{N}\gamma^\mu \left[(\beta_V^0 + \beta_V^1 \tau_3) + (\beta_A^0 + \beta_A^1 \tau_3)\gamma_5 \right] N \,, \tag{5}$$

(V denotes the vector and A the axial-vector component) where $\beta_{V,A}^\tau$ correspond to the couplings for isoscalar ($\tau = 0$) and isovector ($\tau = 1$) interactions, respectively. In models where the $\mu^- \to e^-$ process occurs due to the exchange of scalar particles (exotic Higgs scalars, etc.), [13] the hadronic current is parametrized as

$$J = \bar{N} \left[(\beta_S^0 + \beta_S^1 \tau_3) + (\beta_P^0 + \beta_P^1 \tau_3)\gamma_5 \right] N, \tag{6}$$

Now $\beta_{S,P}^\tau$ involve the couplings for the scalar (S) and pseudo-scalar (P) terms. Obviously, the values of the parameters β_α^τ depend on the specific elementary model assumed for the $\mu^- \to e^-$ process. The effective currents of Eqs. (5) and (6) give rise to both coherent and incoherent contributions.

To obtain the transition width for a particular final state, one usually takes the non-relativistic limit of the above hadronic currents. Then, the partial conversion rate is determined by the square of scalar, vector, etc. matrix elements of the form

$$\mathcal{Q}_\alpha = |\beta_\alpha^0 \mathcal{M}_\alpha^{(0)} + \beta_\alpha^1 \mathcal{M}_\alpha^{(1)}|^2, \qquad \alpha = S, V, A, P \,. \tag{7}$$

If the nucleus of process (1) goes from the initial $|i\rangle$ to a final $|f\rangle$ state, $\mathcal{M}_\alpha^{(\tau)}$

are written as

$$\mathcal{M}_\alpha^{(\tau)} = f_\alpha^\tau \langle f | \sum_{j=1}^{A} \Theta_\alpha^\tau(j) e^{-i\mathbf{q}\cdot\mathbf{r}_j} \, \Phi(\mathbf{r}_j)|i\rangle \,, \tag{8}$$

(f_α^τ denote the known weak form factors approximated by their static values $f_V^0 = 1.0$, $f_A^0 = 1.24$, etc.[10]). $\Phi(\mathbf{r}_j)$ represents the muon wave function and $\mathbf{q}$ is connected to the excitation energy E_x of the nucleus as : $q = m_\mu - \epsilon_b - E_x$, (the nuclear recoil is neglected). m_μ is the mass of the muon, ϵ_b its binding energy in the muonic atom and E_x is referred to the ground state of the daughter nucleus ($E_x = E_f - E_{gs}$).

The functions $\Theta_\alpha^\tau(j)$ contain the spin-isospin dependence of the $\mu^- \to e^-$ operator (the index j runs over all nucleons of the target nucleus). By ignoring the small pseudo-scalar term, the operators $\Theta_\alpha^\tau(j)$ become

$$\Theta_\alpha^\tau(j) = \begin{cases} 1, & \alpha=\text{S,V and } \tau=0 \\ \tau_{3j}, & \alpha=\text{S,V and } \tau=1 \\ \sigma_j/\sqrt{3}, & \alpha=\text{A and } \tau = 0 \\ \tau_{3j}\sigma_j/\sqrt{3}, & \alpha=\text{A and } \tau=1 \end{cases} \tag{9}$$

The evaluation of the quantities $\mathcal{Q}_\alpha$ for every final states $|f\rangle$ in the context of a given nuclear model provides the inclusive $\mu^- \to e^-$ reaction rates. Due to the finite nuclear size, the muon wave function $\Phi(\mathbf{r})$ can be exactly obtained by solving numerically the Shrödinger (or Dirac) equation with the nuclear Coulomb potential. For light nuclei the matrix elements $\mathcal{M}_\alpha^{(\tau)}$ are approximated by factorizing out a suitably averaged muon wave function $\langle \Phi_{1S} \rangle$ and rewriting Eq. (8) as

$$\mathcal{M}_\alpha^{(\tau)} \approx f_\alpha^\tau \langle \Phi_{1S}\rangle\langle f | \sum_{j=1}^{A} \Theta_\alpha^\tau(j) e^{-i\mathbf{q}\cdot\mathbf{r}_j} | i\rangle \equiv f_\alpha^\tau \langle \Phi_{1S}\rangle M_\alpha^{(\tau)} \,. \tag{10}$$

where $M_\alpha^{(\tau)}$ accumulate the pure nuclear physics dependence of the $\mu^- \to e^-$ rates.

The integrals in Eq. (10) are evaluated via the traditional decomposition procedure that leads to the following multipole-expansion operators

$$O_{S,V}^{JM}(\tau) = \delta_{lJ} \sqrt{4\pi} \sum_{i=1}^{A} \theta^\tau(i) \, j_l(qr_i) Y_M^l(\hat{\mathbf{r}}_i), \tag{11}$$

$[\theta^0(i) = 1$ and $\theta^1(i) = \tau_{3i}]$, the Fermi-type components, and

$$O_A^{JM}(\tau) = \sqrt{4\pi} \sum_{i=1}^{A} \theta^\tau(i)\, j_l(qr_i) \left[Y^l(\hat{\mathbf{r}}_i) \otimes \sigma_i / \sqrt{3} \right]_M^J, \qquad (12)$$

the Gamow–Teller-type components. The expression for the branching ratio $R_{\mu e}$ in the general case involves matrix elements of the form of Eq. (7).

3 Nuclear calculations for the inclusive $\mu^- \to e^-$ reaction rates

The inclusive $\mu^- \to e^-$ conversion strengths can be obtained by summing over the partial transition matrix elements for all kinematically accessible final states induced by the relevant multipole operators of Eqs. (11) and (12) as

$$S_\alpha = \sum_f \left(\frac{q_f}{m_\mu}\right)^2 \sum_{JM} |\langle f | \hat{T}_\alpha^{JM} | i \rangle|^2, \quad \alpha = S, V, A \qquad (13)$$

where the operators $\hat{T}_\alpha^{JM}$ are given by

$$\hat{T}_\alpha^{JM} = \sum_\tau \beta_\alpha^\tau\, f_\alpha^\tau\, \hat{O}_\alpha^{JM}(\tau). \qquad (14)$$

The kinematical factor $(q_f/m_\mu)^2$ determines the phase space of the outgoing e^-. The quantities in Eq. (13) can be calculated either with effective nuclear methods (closure approximation, Fermi gas models, etc.) [3] or with detailed calculations performed within nuclear models (QRPA, shell model, etc.).[10,11,12]

4 The coherent $\mu^- \to e^-$ conversion branching ratio

Starting from a general $\mu^- \to e^-$ conversion Lagrangian, it is straightforward to deduce the formula for the branching ratio $R_{\mu e}$. Focusing on the dominant channel of the ground state to ground state transitions we have

$$R_{\mu e}^{coh} = \mathcal{Q}\, \frac{G_F^2\, p_e\, E_e}{2\pi}\, \frac{(\mathcal{M}_p + \mathcal{M}_n)^2}{\Gamma(\mu^- \to capture)}. \qquad (15)$$

The quantity $\mathcal{Q}$ includes mostly scalar and vector strengths [13] of the form of Eq. (7). In the case of R-parity violating interactions, for example, $\mathcal{Q}$ takes the form

$$Q = 2|\alpha_V^{(0)} + \alpha_V^{(3)} \ \phi|^2 + |\alpha_{+S}^{(0)} + \alpha_{+S}^{(3)} \ \phi|^2 + |\alpha_{-S}^{(0)} + \alpha_{-S}^{(3)} \ \phi|^2$$
$$+2 \ \mathrm{Re}\{(\alpha_V^{(0)} + \alpha_V^{(3)} \ \phi)[\alpha_{+S}^{(0)} + \alpha_{-S}^{(0)} + (\alpha_{+S}^{(3)} + \alpha_{-S}^{(3)}) \ \phi]\} \ . \tag{16}$$

The nuclear transition matrix elements $\mathcal{M}_{p,n}$ (proton-neutron representation) of Eq. (15) are defined as

$$\mathcal{M}_{p,n} = 4\pi \int j_0(qr)\Phi(r)\rho_{p,n}(r)r^2 dr \ , \tag{17}$$

(j_0 is the zero order spherical Bessel function) where $\rho_{p,n}(r)$ are spherically symmetric proton (p) and neutron (n) nuclear densities normalized to the atomic number Z and neutron number N of the target nucleus in process (1).

The quantity Q in Eq. (16) depends on the nuclear parameters through the factor $\phi = (\mathcal{M}_p - \mathcal{M}_n)/(\mathcal{M}_p + \mathcal{M}_n)$. For all the experimentally interesting nuclei this parameter is rather small. Therefore, the nuclear dependence of Q can always be neglected except in a very special narrow domain of the parameter space where $\alpha^{(0)} \leq \alpha^{(3)}\phi$.

5 Results and Discussion

The nuclear matrix elements $\mathcal{M}_{p,n}$ are numerically calculated using proton densities ρ_p and neutron densities ρ_n (whenever possible), as in Ref. [13] is described. The muon wave function $\Phi(r)$ is obtained with the Coulomb potential produced by the densities $\rho_{p,n}$ and taking into account the vacuum polarization corrections. The results of $\mathcal{M}_{p,n}$ for the nuclear targets ^{27}Al, ^{48}Ti and ^{208}Pb obtained this way, are shown in Table 1 where other useful quantities (the muon binding energy ϵ_b and the experimental values for the total muon capture rate $\Gamma_{\mu c}$) are also quoted. Using the values of $\mathcal{M}_{p,n}$ and the existing [5] or expected [7,8] experimental limits on $R_{\mu e}$, we can derive upper limits on the LFV parameters entering an effective $\mu^- \rightarrow e^-$ Lagrangian.

The most straightforward limit can be set on the quantity Q of Eq. (15) (see Table 2) but its physical meaning is rather obscure. These bounds were derived from the recent experimental upper bounds on the branching ratio $R_{\mu e}$ of Eq. (3) for the Ti target, and from the expected experimental sensitivity

Nucleus	$q\,(fm^{-1})$	$\epsilon_b\,(MeV)$	$\Gamma_{\mu c}\,(\times 10^6\,s^{-1})$	$\mathcal{M}_p$	$\mathcal{M}_n$
^{27}Al	0.531	-0.470	0.71	0.047	0.045
^{48}Ti	0.529	-1.264	2.60	0.104	0.127
^{208}Pb	0.482	-10.516	13.45	0.414	0.566

Table 1: Transition matrix elements $\mathcal{M}_{p,n}$ (in $fm^{-3/2}$) of Eq. (17) evaluated by using the exact muon wave function obtained via neural networks techniques. [13]

Mechanism	^{27}Al	^{48}Ti	^{208}Pb
$\not{R}_p$ SUSY	$\mathcal{Q} \leq 2.6 \cdot 10^{-19}$	$\mathcal{Q} \leq 0.7 \cdot 10^{-14}$	$\mathcal{Q} \leq 1.1 \cdot 10^{-13}$

Table 2: Experimental limits on the elementary particle quantity $\mathcal{Q}$ defined in Eq. (16).

of the Brookhaven MECO experiment, Eq. (4) for the Al target (for the Pb target we used the limit of Ref. [6]). As can be seen, the limits on $\mathcal{Q}$ for ^{27}Al are improvements by about four orders of magnitude over the existing ones.

In order to obtain upper bounds on other physically more interesting parameters, we adopt the usual assumption that different terms in expression (16) do not substantially compensate each other or, equivalently, that only one term dominates at a time (see Ref. [13]). Such limits can be easily translated into limits on the parameters of a specific model predicting the $\mu^- - e^-$ conversion. Using the upper limits for $\mathcal{Q}$ given in Table 2 we can derive under the above assumptions the constraints on $\alpha_K^{(\tau)}$ of Eq. (16). Thus, the bounds obtained for the scalar current couplings $\alpha_S^{(0)}$ in the R-parity violating Lagrangian for the ^{27}Al target [11] are $|\alpha_S^{(0)}| < 7 \times 10^{-10}$. The limit for $\alpha_S^{(0)}$ obtained with the data for the Ti target [5] is $|\alpha_S^{(0)}| < 1.1 \times 10^{-7}$, i.e. more than two orders of magnitude weaker than the limit of ^{27}Al (see Ref. [13] for limits extracted on other parameters).

6   Conclusions

We developed a systematic approach which allows one to calculate the $\mu^- - e^-$ conversion rates starting from quark level Lagrangians of any elementary model and taking into account the effects of the nucleon and nuclear structure. The transition strengths of the $\mu^- \to e^-$ provide useful nuclear physics input to put severe bounds on the muon number violating parameters determining the effective currents in various models which predict this exotic process.

From the existing data on $R_{\mu e}$ for ^{48}Ti and ^{208}Pb (SINDRUM2 experiment)[5] and the expected sensitivity of the designed MECO experiment[7] we obtained present and expected stringent upper limits on the quantity Q which can, furthermore, be exploited to extract limits on specific LFV parameters.

Acknowledgements

I wish to thank warmly W. Molzon for supporting me to attend the workshop and for his kind hospitality at Honolulu.

References

1. W.J. Marciano and A.I. Sanda, Phys. Rev. Lett. **38** (1977) 1512.
2. T.S. Kosmas, G.K. Leontaris and J.D. Vergados, Prog. Part. Nucl. Phys. **33** (1994) 397; Phys. Lett. **B219** (1989) 457.
3. T.S. Kosmas and J.D. Vergados, Phys.Lett. **B215** (1988)460; *ibid* **B217** (1989)19; Nucl.Phys. **A510** (1990)641; Phys.Rep. **264** (1996)251.
4. T.S. Kosmas Nucl. Phys. **A683** (2001) 443 and references therein.
5. A. van der Schaaf, Prog. Part. Nucl. Phys. **31** (1993) 1 and Private Communication; P. Wintz, Contribution to the present Volume.
6. W. Honecker *et al.*, (SINDRUM2 Coll.), Phys. Rev. Lett. **76** (1996)200.
7. W. Molzon, Springer Tracts in Mod. Phys. **163** (2000) 105.
8. J. Sculli, Contribution to the present Volume.
9. Y. Kuno, Contribution to the present Volume.
10. J. Schwieger *et al.*, Phys. Lett. **B443** (1998) 7; Phys. Rev. **C56** (1997) 2830; T.S. Kosmas *et al.*, Nucl. Phys. **A665** (2000) 183.
11. T. Siiskonen *et al.*, Phys.Rev. **C60** (1999)R62501; *ibid* **62** (2000)35502.
12. T.S. Kosmas *et al.*, Nucl. Phys. **A570** (1994) 637.
13. A. Faessler *et al.*, Nucl. Phys. **B 587** (2000) 25; T.S. Kosmas, S. Kovalenko and I. Schmidt, Phys. Lett. **B**, submitted.

STATUS OF MUON ELECTRON CONVERSION AT PSI

PETER WINTZ

FZ JUELICH, IKP, D-52425 JUELICH, GERMANY

E-mail: p.wintz@fz-juelich.de

Decay modes with a violation of lepton flavor conservation (LFV) test fundamental standard model extensions [1]. The most stringent upper limit on LFV-decay modes comes from the search for muon electron conversion in muonic atoms with the SINDRUM II spectrometer at PSI [2]: $\Gamma(\mu^- Ti \to e^- Ti^{g.s.})/\Gamma(\mu^- Ti \; capture) < 6.1 \times 10^{-13}$ (90% CL). This report summarizes the main SINDRUM II results, searches for μe conversion on different nuclear targets (Ti, Au, Pb) and for charge-changing muon conversion to a positron. Preliminary results of this years μe measurement on gold are reported.

1 Experimental Method

1.1 $\mu \to e$ Conversion in Nuclei

Signature of coherent μe conversion in muonic atoms $\mu^-(A, Z) \to e^-(A, Z)$ is a final single electron with characteristic energy $E_{\mu e} = m_\mu - E_b - E_r$, with E_b muon binding energy and E_r recoil energy of the nucleus (A, Z). The lifetime of muonic atoms of the order of 100 ns is limited by nuclear muon capture $\mu^-(A, Z) \to \nu_\mu(A, Z - 1)$. Up to now no μe conversion has been found resulting in upper limits of the branching ratio $B_{\mu e}$ of μe conversion versus μ-capture rate. Theoretically, $B_{\mu e}$ is predicted[3] to increase with Z and the coherent fraction of the conversion to be larger than 80% for all nuclear systems. Earlier calculations[4] estimated a maximum of $B_{\mu e}$ in the region around $Z = 30$ so that most of the experiments have been performed on medium-heavy nuclei such as Ti. A general model-independent analysis[5] requires measurements on at least two nuclear targets with different values for the normalized neutron excess $(N - Z)/(N + Z)$.

The conversion energy $E_{\mu e}$ coincides with the kinematical endpoint of electrons from bound-μ decay, $\mu^-(A, Z) \to e^- \bar{\nu}_e \nu_\mu(A, Z)$, which is the dominant intrinsic electron background source. The electron energy spectrum has been calculated for a number of muonic atoms [6] and results are shown in Fig. 1. The probability to get an energy within 2% of the endpoint is around 5×10^{-14} resulting in a background well below 10^{-14} per captured μ, which is the proposed μe sensitivity of the SINDRUM II experiment.

Electron background from radiative μ-capture is less significant. Due to the mass difference of initial and final nucleus the kinematical endpoint of the

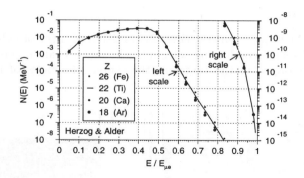

Figure 1. Normalized electron energy spectrum of bound-μ decay for muonic atoms Ar, Ca, and Fe. The spectrum of Ti resulted from these distributions by interpolation.

energy spectrum is lowered and electron contribution from radiative μ capture is around a few percent compared to bound-μ decay.

Non-conservation of total lepton number may lead to the charge-changing conversion $\mu^-(A, Z) \to e^+(A, Z - 2)$. Since the nucleus changes from initial to final state there is no coherent enhancement in this case. Intrinsic positron background comes from radiative μ-capture followed by asymmetrical $e^- e^+$ conversion.

1.2 Prompt and Cosmic Background

Prompt background may be caused by beam e^- scattering off the target and e^- from radiative π^- capture and subsequent pair production. Beam electrons may originate from π^0 decays around the pion production target or from μ^- decays in flight at the end of the beam line. In the first case their momenta reach up to the selected beam momentum. Electrons from μ decay in flight may have significantly higher momenta and to avoid this background completely beam momentum must be below 70 MeV/c.

Negatively charged pions at rest are immediately captured by a nucleus. With a probability of about 1% a photon is emitted with energy above 100 MeV. These photons may convert in the target into asymmetrical $e^+ e^-$ pair and if the e^+ remains unnoticed such events are a background source.

There are numerous processes in which cosmic rays produce e^- in the target with an energy around $E_{\mu e}$. In most of the cases additional particles may be observed, like high-energy μ or associated e^+. Cosmic background can be reduced by passive shielding. A background run with beam switched off allows to study and identify characteristic event topologies.

74

1.3 Spectrometer

The SINDRUM II spectrometer (Fig. 2) consisted of a set of cylindrical detectors inside a superconducting solenoid. The muon beam entered from the left and was stopped in the target at the center. Particles of interest fol-

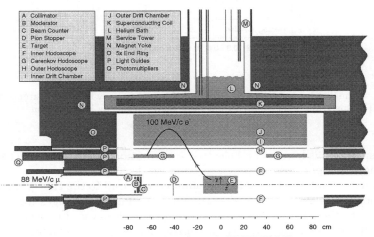

Figure 2. Vertical cross-section through SINDRUM II spectrometer.

low helical trajectories beginning in the target region, making at least one turn before reaching a Čerenkov hodoscope at either end of the spectrometer (Fig. 3). The inner radial drift chamber was used for spacial reconstruction of the trajectories giving the particle momentum. Main purpose of the larger outer chamber was to recognize cosmic ray induced events by additional correlated signals. Scintillator and Čerenkov hodoscope time signals were used to determine particle charge and revolving direction.

Measurements at the $\pi E5$ beam line (see Sect. 2.2) required some upgrades of the spectrometer. A 8.5 m long Pion-Muon-Converter (PMC) solenoid as the last beam line element was connected to the spectrometer. The inner hodoscope was removed and one Čerenkov was replaced by a larger-diameter scintillator hodoscope sitting on the solenoid pipe. The beam vacuum was continued by a low-mass vacuum pipe surrounding the target to avoid muon stops in air and high energy electrons from bound-μ decay in

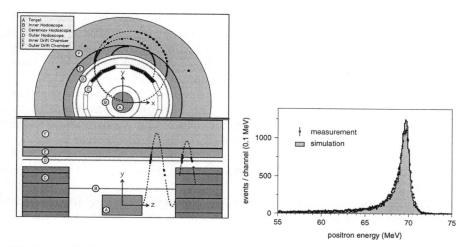

Figure 3. *Left:* Example of a reconstructed electron trajectory in $x-y$ and $y-z$ projection. *Right:* Energy spectrum of positrons from $\pi^+ \to e^+\nu_e$ decay in the low-mass polystyrene target. The measured resolution is 1.5 MeV (FWHM).

muonic oxygen.

To check the momentum calibration and resolution the beam line was tuned to select π^+ which were stopped in a low-mass polystyrene target inside the spectrometer. Fig. 3 shows the measured energy distribution of the monoenergetic positrons from $\pi^+ \to e^+\nu_e$ decay. The magnetic field was reversed and scaled down to the lower momentum compared to μe conversion. The measured spectrometer energy resolution was 1.5 MeV (FWHM) in agreement with simulation. A simulation of μe conversion inside the heavier Ti target yielded a resolution of 2.3 MeV dominated by the spread in energy loss inside the target and sufficient to remain free from bound-μ decay background.

2   Results

2.1   Search for $\mu^- \to e^-$ and $\mu^- \to e^+$ Conversion on Titanium

The 1993 μe measurement [7] was done at the $\mu E1$ beam line. During a live time of 50.4 days with beam switched on 3×10^{13} μ^- stopped in the target with a π/μ contamination around 10^{-7} at the selected beam momentum of

88 MeV/c. Out of 4.0×10^6 recorded events $\simeq 170000$ could be reconstructed as electrons originating in the target. Events caused by radiative π^- capture and scattered beam e^- were removed by a 20 ns wide prompt veto on signals in the beam counter. The large veto width was caused by the flight-time spread of low-energy π^- between second moderator and target.

Cosmic background suppression was checked and optimized by studying data measured with beam switched off. This background was identified by correlated additional signals in the drift chambers and hodoscopes. Cosmic photons entering the spectrometer through the unshielded cryogenic service tower making asymmetric e^+e^- pair production in the target region survived these tests. To remove this background a 3% cut in the geometrical acceptance was applied.

The final electron sample consisted of $\simeq 21000$ events, mainly from bound-μ decay. Radiative μ^- capture and unrecognized scattered beam e^- due to beam counter inefficiency contributed at the level of 1% and 5%, respectively. Fig. 4 shows the electron energy distributions and a μe simulation assuming a branching ratio of $B_{\mu e} = 4 \times 10^{-12}$. The endpoint of the final energy

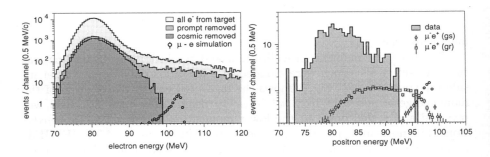

Figure 4. *Left:* Energy distribution of reconstructed electrons from the target and remaining events after suppression of prompt and cosmic ray background. A GEANT simulation of $\mu \to e$ conversion shows the expected amount assuming $B_{\mu e} = 4 \times 10^{-12}$. *Right:* Final positron energy distribution. A simulation of $\mu^- e^+$ conversion with Ca in ground state (gs) and with a giant resonance excitation (gr) shows expected signals using $B_{\mu e}^{gs} = 4.3 \times 10^{-12}$ and $B_{\mu e}^{gr} = 8.9 \times 10^{-11}$.

distribution is at 99 MeV, well below the main μe signal region.

Since no μe conversion candidate was found the measurement resulted in a new upper limit on the branching ratio (with 90% CL) $B_{\mu e} < 6.1 \times 10^{-13}$ which improved on the previous result[8] by a factor of 7.

Since there was no charge decision in the μe trigger also positrons were recorded which allows to search for $\mu^- Ti \rightarrow e^+ Ca$ conversion. The positron energy depends on the mass difference of the nuclei and varies only slightly for the different Ti isotopes if Ca is in ground state. But unlike $\mu \rightarrow e$ conversion, one can expect an excitation of the final nucleus described by a giant dipole resonance with mean energy and width of 20 MeV in case of Ca. This leads to a broad energy spectrum of the positrons and has severe consequences for background discrimination. Fortunately, the intrinsic background is much lower for e^+ than e^-. Fig. 4 shows the energy distribution of the final data sample which consisted mainly of positrons from radiative μ capture and some misidentified scattered beam electrons unrecognized by the beam counter[9].

The endpoint of the positron spectrum was at 92.3 MeV with an isolated event at 95.7 MeV attributed to cosmic ray background in accordance with the measured distribution with beam switched off. Also shown are distributions resulting from $\mu^- \rightarrow e^+$ simulations for ground state and giant resonance final states. Above 92.3 MeV no candidate for $\mu^- \rightarrow e^+$ conversion was found[9] and the new upper limits were (with 90% CL) $B^{gs}_{\mu^- e^+} < 1.7 \times 10^{-12}$ and $B^{gr}_{\mu^- e^+} < 3.6 \times 10^{-11}$ which improved on the previous limits[8] by a factor of 2.5.

2.2 Search for $\mu^- \rightarrow e^-$ Conversion on Gold

A new high-rate pion beam line $\pi E5$ and new Pion-Muon-Converter (PMC) solenoid were build to deliver highest muon flux required to reach the proposed μe sensitivity of a few 10^{-14}. At the expected μ^- and e^- rates a beam counter could not be used anymore. A veto to suppress prompt background would have caused too high losses by accidental beam muons. Therefor, a dedicated pion absorber inside the PMC should yield a negligible pion contamination $\pi/\mu < 10^{-10}$.

Due to cooling problems during the first years the superconducting PMC solenoid could only be operated at limited field strengths making it impossible to inject higher momentum pions with sufficient rate. Nevertheless, in 1997 a search for μe conversion on gold was carried out with a temporarily changed beam concept. The $\pi E5$ channel was tuned to select lower momentum cloud muons of 26 MeV/c for which the PMC field strength was sufficient. At this beam momentum the pion contamination was negligible. A lowest-mass target of $20gr$ was sufficient to stop the beam resulting in a much better energy resolution of $900 keV (FWHM)$ compared to the 1993 measurement and a much lower background scaling with target mass, $e^+ e^-$ conversion from radiative pion, muon capture, cosmic photons and target scattering background.

During 13.5 days 8×10^{11} muons were stopped. The final data consisted

78

of electrons from bound-μ decay with endpoint at 88 MeV/c, well below an expected μe-signal around 95 MeV/c. No evidence for μe-conversion was found, resulting in a new upper limit [10] (with 90% CL) of $B_{\mu e} < 2 \times 10^{-11}$ with a factor 2.5 improvement on our limit for μe-conversion on lead [11].

In 1999 a μe-measurement on Ti was done with a factor 3 higher muon stop rate compared to 1993. Data of the first several weeks were checked on-line to be background-free giving us the possibility to increase muon stops by slightly raising beam momentum (with adjusted moderator thickness). But offline data analysis showed we raised background from radiative pion capture drastically from less than 10^{-4} up to one per minute making it impossible to search for μe-conversion in the data.

In 2000 we started a new μe-measurement on gold beginning with several weeks of systematic pion beam contamination studies. A beam line setting (of slits, focusing and bending magnets) was found cutting the pion high momentum tail by some orders of magnitudes but only loosing about 25% of muon rate. Background rate from radiative pion capture in the gold target was determined for different beam momenta using the full event reconstruction analysis and looking for high momentum e^- and e^+ target events.

For the μe measurement a momentum of 50 MeV/c was selected. During 1500 h around 3×10^{13} muons were stopped in a 50 gr Au target yielding a μe-sensitivity around 2×10^{-13} (no signal, no background, 16% μe efficiency assumed). Fig. 5 shows the preliminary final electron spectrum below the μe-conversion region around 95 MeV, after applying cuts to remove cosmic and bad-reconstruction background and a geometrical cut on the target. Mainly electrons from bound-μ decay remain. Momentum resolution was checked to be around 1.2 MeV/c by measuring the sharp energy edge at 52 MeV/c of decay e^+ from μ^+ stopped in the target.

Acknowledgments

First thanks to W. Molzon and Y. Kuno for invitation and organisation of this interesting workshop. Special thank to A. van der Schaaf for discussions about latest results.

References

1. J.Feng, Y.Okada, K.Tobe, T.S.Kosmas, A.Czarnecki, W. Marciano, these proceedings.
2. PSI proposal R-87-03 (1987), A. van der Schaaf (spokesman).
3. H.C. Chiang et al., Nucl. Phys. A **559**, 526 (1993), T.S. Kosmas et al., Phys. Rev. C **56**, 526 (1997).

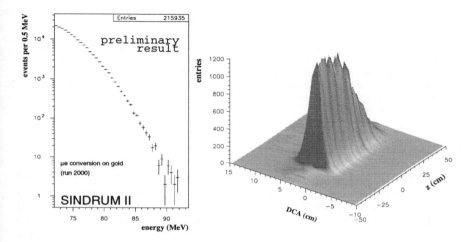

Figure 5. *Left:* Energy spectrum of electrons surviving cuts to remove cosmic and bad-reconstruction background. *Right:* Target distribution of reconstructed electrons, radial distance (DCA) and z-coordinate of closest target approach.

4. S. Weinberg and G. Feinberg, *Phys. Rev. Lett.* **3**, 111 (1959), T.S. Kosmas and J.D. Vergados, *Phys. Lett.* B **217**, 19 (1989), T.S. Kosmas and J.D. Vergados, *Nucl. Phys.* A **510**, 641 (1990).
5. O. Shanker, *Phys. Rev.* D **20**, 1608 (1979).
6. F. Herzog and K. Alder, *Helv. Phys. Acta* **53**, 53 (1980), R. Watanabe *et al.*, Atomic Data and Nuclear Data Tables 54, 165 (1993), O. Shanker and R. Roy, *Phys. Rev.* D **55**, 7307 (1997).
7. P.Wintz, Proceedings of the 29th International Conference on High Energy Physics, ed. A.Astbury, D.Axen, J.Robinson, World Scientific, Singapore (1999). P. Wintz, *Ph.D. thesis*, RWTH Aachen (1995). S. Eggli, *Ph.D. thesis*, University of Zurich (1995).
8. SINDRUM II Collab., C. Dohmen *et al.*, *Phys. Lett.* B **317**, 631 (1993).
9. SINDRUM II Collab., J. Kaulard *et al.*, *Phys. Lett.* B **422**, 334 (1998).
10. F. Riepenhausen, *Ph.D. thesis*, University of Zurich (2000).
11. SINDRUM II Collab., W. Honecker *et al.*, *Phys. Rev. Lett.* **76**, 200 (1996).
12. G.Kurz, F.Rosenbaum, *Ph.D. theses*, University of Zurich, J.Kuth, *Ph.D. thesis*, RWTH Aachen, all to be submitted.

THE MECO EXPERIMENT

JOHN SCULLI

New York University

An experiment is described to search for muon to electron conversion in the field of a nucleus with one event sensitivity, normalized to the muon capture rate on the same nucleus, of 2×10^{-17}. The muon to electron transition does not conserve lepton flavor. For it to occur at an observable rate new physics is required, beyond the usual Standard Model and minimal extensions that include the possibility of light massive neutrinos.

1 Introduction

The interest in muon to electron conversion as a probe of physics beyond the Standard Model is discussed in a number of the contributions to this Workshop. This presentation is devoted to describing the proposed MECO experiment, [1] and the methods by which it intends to improve upon previous searches. The process sought is one in which a muon brought to rest in matter converts to an electron coherently in the field of a nucleus,

$$\mu^- + (Z, A) \longrightarrow e^- + (Z, A). \tag{1}$$

The electron's energy is very close to the rest energy of the muon,

$$E_e = E_\mu - \frac{E_\mu^2}{2M_A}, \tag{2}$$

where E_μ is the muon energy before capture, mass plus binding energy, and M_A is the nuclear mass. An electron of this energy, detected in a time window delayed with respect to the muon stop, signals the conversion. In the free decay of a muon at rest to an electron and two neutrinos, the electron's energy is at most half the muon rest energy. In the decay of a bound muon, however, the energy approaches that of the conversion electron when the two neutrinos carry away little energy. In this limit, the electron recoils against the nucleus only, mimicking the two body process that distinguishes muon to electron conversion. The spectrum falls rapidly near the kinematic limit, as $(E_{max} - E)^5$, and the probability that the decay electron is within 3 MeV of the endpoint is 5×10^{-15} in aluminum, the material of choice in MECO.

The distinctive signature of the $\mu \to e$ transition has made possible a number of sensitive searches, the list dating back to the 1955 experiment of Steinberger and Wolfe, [2] with experiments of ever increasing sensitivity in the decades since. The two most recent are the 1988 TRIUMF experiment [3] that reached a sensitivity of R < 4.6×10^{-12} and the 1993 SINDRUM2 experiment [4] at the PSI that achieved a limit R < 6.1×10^{-13}, where R measures the rate compared to the muon capture rate in the same nucleus:

$$R = \frac{\mu^- + (Z, A) \longrightarrow e^- + (Z, A)}{\mu^- + (Z, A) \longrightarrow \nu + (Z - 1, A)}. \tag{3}$$

These sensitivities were achieved in background free experiments, with no event candidates in a suitably chosen window about the expected conversion energy. The durations of these experiments, typically ≤ 1 yr ($\sim 10^7$ seconds), and beam intensities, 10^6 stopped muons per second, lead to the quoted sensitivities when proper account is taken of the detection efficiencies. To achieve the 2×10^{-17} MECO sensitivity in an experiment of comparable duration and efficiency $\sim 10\%$, requires a beam of 10^{11} muons/sec and background four orders of magnitude below the level required in these earlier experiments.

2 MECO Features

The design features that make possible the required large muon flux and the necessary background rejection are described here briefly. The experiment is built around three superconducting solenoids and uses a pulsed 8 GeV/c proton beam at the Brookhaven National Laboratory's Alternating Gradient Synchrotron.

Production Solenoid

In Fig. 1 the MECO experimental arrangement is shown and the three solenoidal magnets are identified. The field in the *production solenoid* varies from 2.3 Tesla at the location of the target to 5 Tesla at its end. The proton beam enters the solenoid moving in the direction of increasing field, opposite the outgoing muon beam direction and away from the detectors. Pions and decay muons moving in the forward direction but outside the graded magnet's loss cone (of half angle $\sim 43^\circ$) are reflected by the higher field and, following helical trajectories, join the backward moving pions and muons. Those with less than 180 MeV/c are confined within the 30 cm inner radius of the magnet's shielding. A large fraction of the confined pions decay, producing muons that leave the low field region and enter the transport solenoid. The arrangement yields 0.0025 stopped muons/proton, or 10^{11} muons/sec for the anticipated 4×10^{13} protons/sec on target. [5]

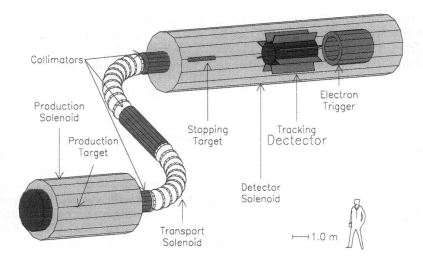

Figure 1. The production, transport, and detector superconducting solenoids of the proposed MECO Experiment.

Curved Transport Solenoid

Muons are transported to the muon (stopping) target by the *transport solenoid*, which also serves to reduce the number of unwanted particles made in the production solenoid that reach the detectors. The transport solenoid displaces the production solenoid ~ 13m from the production target and provides momentum selection through the vertical curvature drift in its two bent sections. Beam μ^-'s with momentum less than 60 MeV/c reach the target while μ^-'s of greater momentum, which would pass through the target, are absorbed in the collimators. Beam μ^+'s and electrons of momentum greater than 100 MeV/c are not transported. Antiprotons absorbed anywhere in the vicinity of the muon target are a potential source of background because of their long drift time down the solenoid. They are absorbed in a thin Be window at the central collimator located between the two curved sections of the solenoid.

Detector Solenoid

The muon target and the detectors are located in the *detector solenoid*. The target is centered in a graded field that varies from 2T at the upstream end, closest to the transport solenoid, to 1T approximately 4 meters downstream. The tracking detector is not centered on the target but displaced downstream by 4 meters. The detector region is shown in Fig. 2 and the straw tube tracker and trigger calorimeter schematically in Fig. 5. Conversion electrons produced ±30° about 90° are accepted, the backward moving electrons reflected in the graded field. The minimum P_T is 90 MeV/c at production and the solid angle acceptance is 1/2. The electrons move in the graded field with P_T^2/B constant and reach the detector with 75 MeV/c < P_T < 86 MeV/c, moving in a helical trajectory that passes through the target and has radius greater than 25 cm. Decay electrons, with P_T < 53 MeV/c pass down the center of the solenoid and do not intercept the tracker or calorimeter. Another feature of this arrangement is that electrons made upstream with P = 105 MeV/c have a maximum P_T of 75 MeV/c after passing through the field gradient; they are eliminated by a P_T cut unless they scatter in the target. The displacement of the detectors downstream of the muon target reduces the rate from neutrons and gammas.

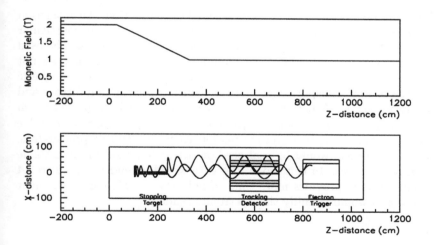

Figure 2. Top: The magnetic field in the detector solenoid. Bottom: Two conversion electrons that reach the detectors, one produced in the forward direction and the other produced backward and reversed in the graded field.

Pulsed Beam

While it is unlikely that a muon stopping in the target will result in a decay electron close to the conversion energy of 105 MeV, a pion can readily produce such an electron. For example, in the familiar process $\pi^- + p \to n + \gamma$, the photon has energy 130 MeV and can convert asymmetrically, either internally or through a secondary interaction, to produce an electron with energy 105 MeV. The distinction is that these electrons are produced promptly, coincident in time with the stopping pion, rather than delayed by the characteristic muon lifetime, 880 nsecs in aluminum. To reject this and other prompt backgrounds, the AGS proton beam RF structure is preserved on extraction. Two possible timing schemes are being considered. In the first, depicted in Fig. 3, pulses of duration $\leq$ 100 nsecs separated by 1.35 μsecs are each filled to an intensity of $\simeq$ 2×10^{13} protons/sec. This is accomplished by filling two of six equally spaced RF buckets that occupy the 2.7 μsec revolution time around the ring. The gate for detecting the conversion electron extends from 600 to 1250 nsecs, with 100 nsec available to hold the signals before the next beam pulse. During the gate, there are no beam particles and, ideally, no pions stopping in the target, thus eliminating the prompt background. The ratio of the beam between the filled buckets to the beam in the buckets is called the extinction. An extinction of 10^{-9} is required. Values of 10^{-7} have been measured with no particular tuning of the AGS pulsed beam. A kicker magnet in the AGS ring will be used to reduce the extinction to the desired value if beam tuning alone does not lower it the additional two orders of magnitude required. A pulsed magnetic kicker in the proton transport is planned to reduce the extinction further and to measure its value. A better mode of operation might be a pulsed beam with 2.7 μsec spacing. This could be achieved by running the AGS with twelve buckets in the ring, filling two adjacent buckets each with 2×10^{13} protons/sec. The two filled buckets would be coalesced on extraction resulting in a single bunch every 2.7 μsecs. In this mode, the gate would extend from 0.9 to 2.4 μsecs. This scheme provides more time to hold the signals before the next beam pulse and would simplify the electronics readout.

Detectors

The helical trajectory of the electron is measured in a straw tube tracker that consists of 16 identical planar detectors, 0.3 m across and 2.4-2.9 m along the axis of the solenoid. Each planar detector has 3 layers of straws, 2880 total in the 16 planes. The resolution is 1 MeV fwhm and is dominated by multiple scattering, with a low energy tail from energy loss in the target. The resolution function is shown in Fig. 4 with background from muon decay in

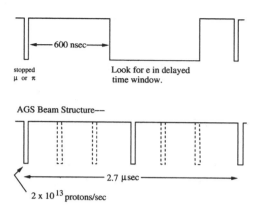

Figure 3. Prompt background is rejected by exploiting AGS Beam Structure. In one scheme, two of six buckets in the ring are filled and conversion electrons sought in the 1.35 μsec region between the filled buckets, where the beam rate is less by $\times 10^{-9}$.

orbit superimposed. At a length of 2.9 m, the tracker records two full turns on the trajectory. Every legitimate (x, y, θ) cluster, where θ is the local angle measured by the three straw planes, would then have a downstream clone. This feature reduces the resolution tails produced by background from extra hits in the tracker, for example, events in which a decay in orbit electron's momentum is mismeasured because of nearby random hits.

The event is triggered by energy deposition in a crystal calorimeter with good energy ($\sigma \leq 5\,\text{MeV}$) and spatial ($\sigma \sim 1\,\text{cm}$) resolution. An energy correlated hit on the electron's trajectory reduces combinatoric background further. The good energy resolution permits a high energy threshold that limits the number of decay in orbit triggers that must be fully reconstructed. A 2000 element detector, e.g. 3 x 3 x 12 cm lead tungstate crystals arranged in four bars each 150 x 30 x 12 cm, provides good acceptance. An event in the proposed tracker/calorimeter detector, viewed along the solenoid axis, is shown in Fig. 5.

3 Summary

Compared to earlier experiments, the proposed MECO experiment is a substantially scaled up search for muon to electron conversion. It will observe 5 events at R = 10^{-16} for 10^7 seconds of data taking. Prompt background is suppressed by using a pulsed beam from the AGS. Studies using GEANT

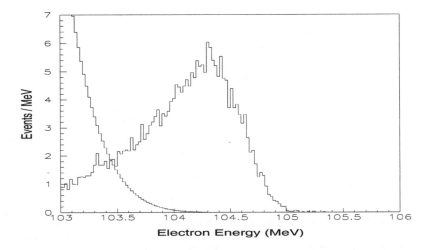

Figure 4. GEANT simulation of signal and muon decay in orbit in the MECO detector. For $R = 10^{-16} \sim 5$ events are expected above 103.6 MeV in 10^7 seconds of data taking.

indicate that muon decay in orbit will be the dominant background, as in earlier experiments. These studies include background from reconstruction errors caused by extra hits in the detector.

Acknowledgments

Research supported by the National Science Foundation.

References

1. The MECO Collaboration, in *Rare Symmetry Violating Processes, A Proposal to the National Science Foundation to Construct the MECO and KOPIO Experiments*, October (1999). Available from http://meco.ps.uci.edu/RSVP.html.
2. J. Steinberger H.B. Wolfe, *Phys. Rev.* **100**, 1490 (1955).
3. S. Ahmad *et al.*,TRIUMF, *Phys. Rev.* D **38**, 2102 (1988).
4. C. Dohmen *et al.*, SINDRUM2, *Phys. Lett.* B **317**, 631 (1993), and P. Wintz, these Proceedings.
5. See V. Tumakov, these Proceedings.

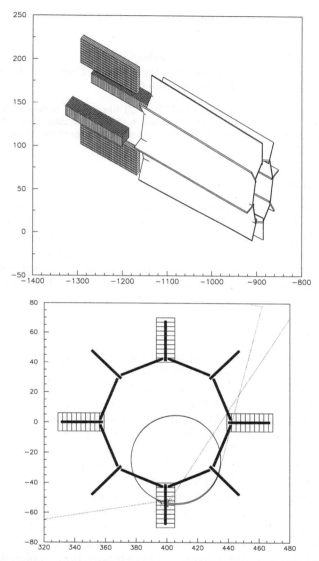

Figure 5. Top: The proposed MECO detector. A 2.4-2.9 m straw tube detector with eight planar elements is followed by a 2000 element crystal calorimeter. Bottom: A GEANT event viewed along the axis of the solenoid. A conversion electron (circle) produces four hits per turn in the tracker before being absorbed in the calorimeter. Straight lines leaving calorimeter are low energy photons.

High Intensity Muon Beam for $\mu - e$ Conversion Experiment

V.L.Tumakov[a]

Department of Physics and Astronomy, University of California, Irvine, USA

E-mail: tumakov@uci.edu

The results of simulations based on GEANT 3.21 library and technical details of the proposed MECO experiment are discussed in this paper: time structure of the proton beam, experimental setup and fluxes of secondary particles at entrance of the detector solenoid.

1 Introduction

The MECO experiment is proposed to search for the rare lepton flavor violation process $\mu^- N \rightarrow e^- N$ with far greater sensitivity than in any past experiment [1-3]. The experiment requires a very intense low energy μ^- beam. The idea is to capture the muons from the proton interaction in the magnetic field of the production solenoid and then transport with predominant selection low energy negative muons by means of curved solenoid to the stopping target in the detector solenoid.

2 Proton beam

A very intense pulsed muon beam is critical to MECO, both to achieve the desired sensitivity and to reduce background.

The Alternating Gradient Synchrotron (AGS) at the Brookhaven National Laboratory has operated with 6,8, and 12 buckets in the 2.7 μs revolution time. It is proposed to operate the AGS with the harmonic number 6 (6 RF buckets in the revolution time) and fill only two RF buckets separated by half of the circumference, t= 1.35 μs. Figure 1 shows the times required for various stages of the AGS operation in the first case. The time structure of the proposed beam is shown schematically in Figure 2.

It is very critical for the background level to achieve a low value ($< 10^{-9}$) of the extinction (extinction is the ratio of total beam between the filled buckets to the beam in the filled buckets). Special tests of the bunched beam properties have been done [4,5]. The measured extinction is in the range $0.8 - 1.25 \times 10^{-7}$.

Two possibilities to improve the extinction have been explored. The first involves a system of kickers in the AGS ring. This method of improving the extinction has the advantage that the kickers will run continuously during acceleration and require relatively small field since the beam is kicked many

---

[a]representing MECO collaboration [1]

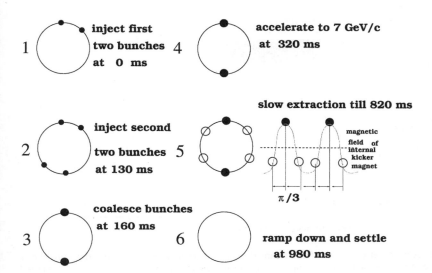

Figure 1: The scheme shows the times required for the various stages in injection, acceleration, extraction for the MECO operation mode of the AGS, and also the operation diagram for external kicker.

times. The basic idea of the system is to use two magnets in the AGS ring. One magnet produces a field modulated at 60 kHz. This would serve to destabilize the beam, and only low field is required for this purpose. To preserve the stability of the beam in the filled RF buckets, a kicker is operated at the frequency of the filled RF buckets, about 740 kHz in the case of two filled buckets in the 2.7 μs revolution time of the machine. The field integral in this kicker is adjusted to be equal and opposite in magnitude to that of the sinusoidally modulated magnet, and it fires only when the filled buckets pass through it. Hence, the net momentum transfer to protons in the filled RF buckets is zero. A second solution is an external kicker in the proton transport line modulated with sinusoidal field. It can be seen from the diagram in Figure 1 that in this scheme, the momentum transfer for filled and unfilled buckets is opposite.

To provide a very intense muon beam it is proposed to operate the AGS with a total of 4×10^{13} protons per cycle (2×10^{13} protons in each filled RF bucket). Currently, the maximum intensity that has been demonstrated is $\sim 10^{13}$ protons per RF bucket. The optimism to achieve the required intensity is based on two differences in MECO vs. standard running conditions. First, only two transfers from the booster to the AGS will be required. Hence, beam will be stored at transfer energy, where space charge effects are most severe,

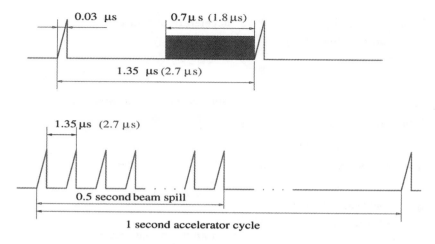

Figure 2: A schematic picture of the beam time structure is shown. The top drawing shows the micro-structure, with 30 ns long proton pulses separated by 1.35 (2.7) μs. The shaded region is the time during which conversion electrons are detected. The bottom drawing shows the macro-structure with a 0.5 s long train of pulses separated by 1.35 (2.7) μs in a 1.0 second long accelerator cycle.

for only 160 ms. Second, the beam will not be accelerated through transition ($\sim$ 8 GeV). Beam instabilities at transition restrict the bucket density during normal operations and this limitation will not exist.

No tests have yet been done of operation with the desired intensity, but it is expected that after some modifications the AGS accelerator can satisfy the requirements of the MECO experiment.

3   Muon production region

The current design of the super-conducting production solenoid has maximum value of the axial component of the magnetic field 5 T near the upstream end of the solenoid, decreasing linearly to 2.3 T at the downstream end, shown in Figure 3. Generated muons will be reflected in the graded field if their initial momentum has direction opposite to that of the entrance to the transport solenoid, resulting in good acceptance.

In comparison to the proposed 1-4 MW beam in different projects of neutrino factory [6-8], the problems in the production region of 60kW AGS proton beam should be less. The optimization must solve two major problems with a high intensity beam: cooling of the production target and shielding of the solenoid's cold mass from the heat load due to radiation.

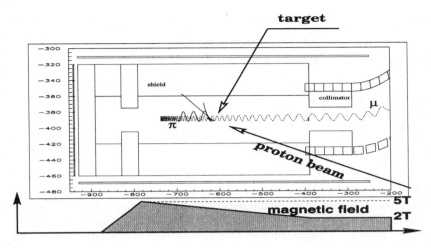

Figure 3: GEANT simulation of the muon production. Axial component of the magnetic field as a function of Z coordinate is plotted below the schematic view of the production region.

3.1 Target cooling

Tungsten's high melting temperature allows the use of a radiation cooled target. The calculations show that the target temperature stops rising after 25 seconds and than fluctuates about the mean value (2660 K) by ±60 K. Even though the temperature is well below Tungsten's melting point, material will evaporate from the surface. At ~ 2650 K the evaporation rate is ~ 10^{-8}g cm^{-2} s^{-1}.

To decrease the mechanical stresses in the target we explored the possibility to divide the target on separated disks, to use high emissivity (0.9) Rhenium coating and studied the advantages of conical shape of the target with wide spread (≈0.3 cm) of the proton beam on the target.

3.2 Solenoid heat load

The heat and radiation load from the particle spray on the super-conducting solenoid surrounding the production target could cause the magnet to quench or fail due to radiation damage, and in any case will represent a heat load on the refrigeration system. Simulations using GEANT have show that a combination of copper and/or Tungsten shielding in a cylindrical shell surrounding the 30 cm radius clear bore can reduce the local instantaneous heat load, the average heat load, and the radiation load integrated over the lifetime of the

Table 1: The table gives the energy deposited in a model of the production solenoid cold mass and radiation load during running time $10^7 s$ for different heat shield configurations. In all cases, the shield has an internal radius of 30 cm and a length of 440 cm.

Configuration	Average total power, W	maximum power, $\mu W/g$	max. radiation load, MRad
30 cm Copper	108	151	146
40 cm Copper	52	65	62
30 cm Tungsten	28	43	41
40 cm Tungsten	10	14	14

experiment to a manageable level.

To estimate the heat load on the super-conducting coils, a GEANT simulation was run for 8 GeV protons hitting the Tungsten target inside the super-conducting solenoid[9]. Copper and/or Tungsten shields of different thicknesses were studied. In preliminary calculations, the solenoid cold mass was approximated by a 6 cm thick aluminum shell immediately outside the shield.

Table 1 gives results for the total heat load, the maximum instantaneous local heat load and the maximum radiation load in the lifetime of the experiment. Even for the case of a shield of 30 cm of copper, all three parameters are acceptable from the point of view of reliable operation of the magnet and longterm radiation damage. We anticipate that we will use a mostly copper shield with some heavy inserts in the region of most intense particle spray in order to reduce the heat load to below 50 W.

4 Transport solenoid

Muons are transported from the production solenoid to the detector solenoid using a curved solenoid bent first by 90° in one direction, and then 90° in the opposite direction, as shown in Figure 4a. The purpose of the bends is to decrease the transmission of both high momentum particles and positive particles. Unwanted particles are absorbed in appropriately shaped collimators at the ends of the transport and at the center.

The simulation of the beam transport is based on an initial design[10] of the solenoid that used 54 "coil packs" to produce a field with axial component of approximately 2 T. The packs are arranged in the curved configuration with appropriate gaps for mechanical considerations. The magnetic field was calculated exactly using the law of Biot and Savart, then used in GEANT simulations to accurately integrate the trajectories including effects of field inhomogeneities.

Charged particles of sufficiently low momentum follow helical trajectories centered on magnetic field lines. In a torus, they drift in a direction perpendicular to the plane of the torus, by an amount given by $D_y = \frac{1}{0.3B} \times \frac{s}{R} \times \frac{p_s^2 + \frac{1}{2}p_t^2}{p_s}$, where D_y is the drift distance, B is the magnetic field, s/R is the bend angle of the solenoid, and p_t and p_s are the perpendicular and parallel momentum components. For $s/R = \pi/2, p_t = 0.09$ GeV/c, $p_s = 0.12$ GeV/c, and $B = 2$ T, the drift of the center of the helix is 49 cm. The drift direction depends on the charge. Hence, by putting appropriate collimators in the straight sections, negative particles of high momentum can be absorbed and flux of positive particles can be suppressed. The drift effect in the trajectory is illustrated in Figure 4b.

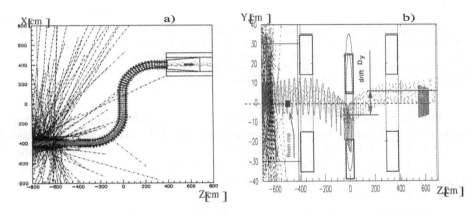

Figure 4: The schematic view of the transport solenoid in two projections in a GEANT simulation. The plot (b) shows a few simulated trajectories in the transport solenoid. A downward drift D_y in its first curved portion (Z=-400 to 0 in the figure) is apparant; the amount of drift depends on the momentum. Both particles are negative.

5 Beam composition at the entrance of the detector region

Figure 5 shows the fluxes of μ^- in comparison with fluxes of the background electrons, positrons and μ^+ at the exit of the transport solenoid. The time (left) and momentum (right) distributions are shown.

The net yield of μ^- at the entrance to the detector solenoid is about 0.005 μ^- per proton with stopping probability in the target 56%, and relative fluxes of the μ^-, e^-, μ^+, e^+ particles are correspondingly 1 : 12 : 0.06 : 2.1.

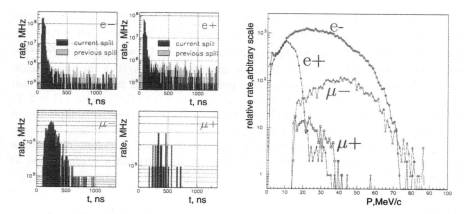

Figure 5: Time and momentum distributions for particles that exit the transport solenoid. The plots on the left show the time distribution, folded about the pulse spacing; the plot on the right shows the momentum distribution of each particle type. The effect of the asymmetric collimator on the flux of negative and positive particles is apparent, as is the effect on the momentum distribution, particularly for positives.

The fluence of electrons is very important, as it is a potential cause of the background. Electrons form the major component of the beam, and represent a significant contribution to detector rates. We have minimized the electron flux by extending vacuum farther upstream of the target to a low field region and by placing a small beam dump downstream of the target to absorb low energy electrons and positrons.

Low momentum anti-protons (< 100 MeV/c) can propagate down the transport solenoid; they have very low kinetic energy and velocity and will take a very long time to transit the transport solenoid. For this reason, they are not suppressed by the beam extinction and represent a potential problem. It is found [11] that a thin Be absorber in the transport system reduces the background, induced by $\bar{p}$'s to negligible level, 10^{-3} events/10^7s.

Pions arriving at the stopping target in the detector solenoid are immediately captured by a nucleus after they stop, with some probability to give 105 MeV electrons by conversions of γ's from radiative capture, so pions represent potential problem for MECO experiment. Simulations show that the contribution to the background from late arriving pions in the accepted time window (700 ns after the proton pulse) is 0.001 events in the experiment lifetime.

6 Summary

After some modifications, the AGS accelerator will satisfy the requirements of the proton beam for the MECO experiment: high intensity, proper time structure and a high extinction. Simulations based on GEANT package show acceptable levels of heat and radiation load on the cold mass of the production solenoid and the Tungsten production target. The graded magnetic field of the production solenoid provides a high acceptance for muons. The curved transport solenoid provides good rejection of positive particles and of high momentum particles and a negligible level of π^- and $\bar{p}$ background in the detector region.

References

1. W.R.Molzon *et al.*, proposal to the National Science Foundation, http://meco.ps.uci.edu/RSVP.html.
2. J.Sculli, report on this conference.
3. W.R.Molzon, proceedings of International Symposium on Lepton and Baryon Number Violation, Trento, Italy, 1998.
4. R.Lee, note meco017, http://meco.ps.uci.edu/ (unpublished).
5. J.Sculli, note meco039, http://meco.ps.uci.edu/ (unpublished).
6. R.Palmer, Report on this conference.
7. A.Riche, Report on this conference.
8. Y.Mori, Report on this conference.
9. D. Garcia and W.R.Molzon, note meco018, http://meco.ps.uci.edu/ (unpublished).
10. W.R.Molzon, note meco008, http://meco.ps.uci.edu/ (unpublished).
11. R.Lee, note meco026, http://meco.ps.uci.edu/ (unpublished).

PRISM

YOSHITAKA KUNO

Department of Physics, Graduate School of Science,
Osaka University,
Machikane-yama 1-1, Toyonaka, Osaka, Japan 560-0043
E-mail: kuno@phys.sci.osaka-u.ac.jp

A Japanese PRISM project which is aiming to construct a highly-intense slow muon source with narrow energy spread and high purity is described.

1 Introduction

There is a strong demand to construct a high intensity muon source, and it is growing rapidly. One of the physics motivations for highly-intense muon source in the field of high energy physics is a search for muon lepton flavor violation (LFV)[1]. It is known that to make substantial improvements in a sensitivity of the LFV searches in future, a high-intensity muon beam is mandatory.

The idea of realistic realization of a high-intensity muon source has emerged in various studies of a $\mu^+\mu^-$ collider at the high-energy frontier and also a neutrino factory based on a muon storage ring. Low-energy muons available from the front-end of those complexes would provide great opportunity of muon physics with higher precision.

Most of the muon LFV experiments use a stopped muon beam. Therefore, the requirements necessary to pursue significant progress would be (1) a narrow energy spread in a muon beam (to increase a stopping efficiency and improve detector resolution for the daughter particles from rare muon decays), and (2) low contamination of a muon beam (to reduce backgrounds, in particular pions in a beam). There is a project which is aiming to construct such a muon source. It is called the PRISM project. In this note, the PRISM project will be introduced. In the following section, the overview of the PRISM project is shown. The pion capture and the phase rotation sections are described in Section 3 and 4, respectively. The application of PRISM other than the search for LFV will be described in Section 5.

2 What is PRISM ?

PRISM is the project in Japan to make a dedicated source of a high intensity muon beam with a narrow energy-spread and less contamination [2,3]. PRISM stands for a "Phase-Rotated Intense Slow Muon source". The aimed intensity is $10^{11} - 10^{12} \mu^{\pm}$ /sec, four orders of magnitude higher than that available at present. It is achieved by a large solid-angle pion capture with a high solenoid magnetic field. The narrow energy spread can be achieved by phase rotation which accelerates slow muons and decelerates fast muons by an radio-frequency (rf) field. The pion contamination in a muon beam can be removed by a long flight length of a beam so that most of pions decay out. Therefore, PRISM would consist of

- a pulsed proton beam (to produce a short pion pulsed beam),

- large-acceptance pion capture by high solenoid field,

- pion decay section (in a long solenoid magnet of about 10 m long), and

- phase rotation section (which accelerates slow muons and decelerates fast muons by a high-gradient rf field).

Some of the key components are explained in detail in the following sections. The conceptual structure of PRISM is shown in Fig.1. The expected PRISM beam characteristics is summarized in Table 1.

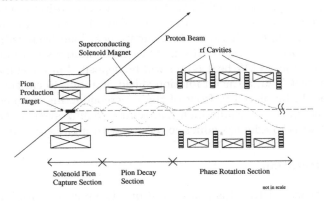

Figure 1. Conceptual Components of PRISM. It has the pion capture, the pion-muon decay section and the phase rotation.

3 Solenoid pion capture

In pion capture, low-energy pions (and muons) are trapped in a high solenoid magnetic field. The magnetic field is 10 T or more. When a magnetic field is

Table 1. Expected PRISM Beam Design Characteristics

Parameters	design goal	comments
Intensity	$10^{11} - 10^{12}$ $\mu^{\pm}$/sec	
Muon Kinetic Energy	20 MeV	$P_{\mu} = 68$ MeV/c
Kinetic Energy Spread	$\pm(0.5 - 1.0)$ MeV	
Beam Repetition	1 kHz	increased if technically feasible.

higher, the beam emittance of muons becomes better and smaller. Therefore, the higher magnetic field is preferable. To estimate a rate of captured pions, Monte Carlo simulations have been made based on GEANT with fluka or MARS hadron production codes. Typical pion spectra which are obtained from Monte Carlo simulation from a 50-GeV proton beam incident are shown in Fig.2.

From Fig.2, about 0.3 captured pions (of momentum less than 0.1 GeV/c) per proton are expected for a proton beam energy of 50 GeV. For a proton intensity of about 10^{14} protons/sec which is planned in the 50-GeV proton synchrotron (PS) in the KEK/JAERI Joint Project[4], captured pions of about 3×10^{13}/sec are expected, and they are sufficient for the aimed intensity.

To realize a high solenoid field at the proton target position, several superconducting solenoid magnets have to be installed. However, they are expected to be exposed to high radiations. To estimate heat load from radiation (primarily from neutrons), the MARS simulation calculations were done. They showed a few hundred kW heat power will be deposited in those magnets. To reduce such heat load, a thick radiation shield of about 40 cm thick made of tungsten should be placed in front of the superconducting magnet. In addition, it is necessary to include high and robust cooling system. One idea of cooling system is to adopt conductive cooling with high heat transfer path.

4 Phase Rotation

The key element of PRISM is phase rotation. The phase rotation is to accelerate slow muons and to decelerate fast muons by a strong radio-frequency (RF) field, yielding a narrow longitudinal momentum spread. It corresponds rotation of the muon phase space in the energy-time space, as shown in Fig.3. After phase rotation, the projection of the phase space onto the axis of energy becomes narrower and sharper. To identify fast and slow muons, a time of flight (ToF) from the time of production of their parent pions is used. Therefore, a very narrow pulsed proton beam must be used to determine their ToF precisely.

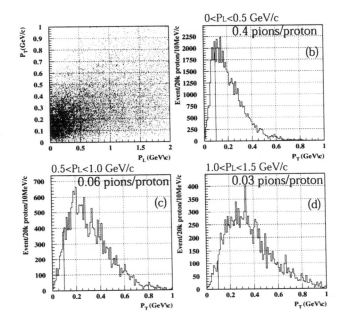

Figure 2. Pion Spectra produced by 50-GeV protons. (a) P_t (transverse momentum) vs. P_L (longitudinal momentum), (b) P_T spectrum for $P_L < 0.5$ GeV/c, (c) P_T spectrum for $0.5 < P_L < 1.5$ GeV/c, (d) P_T spectrum for $1.0 < P_L < 1.5$ GeV/c

4.1 FFAG for phase rotation

One of the features of the phase-rotation section in PRISM is to adopt a circular system based on a Fixed-Field Alternating Gradient synchrotron (FFAG), instead of a linear system (linac).

The circular system is much better than the linear system since the full size is smaller and more compact, and the number of rf cavities and powers needed is smaller. Therefore a total cost is smaller.

Among the other circular machines (such as synchroton and cyclotron), FFAG would have several advantages: FFAG has a large momentum acceptance and large transverse emittance, whereas a synchrotron has very small acceptance and a cyclotron that could have a large acceptance does not have

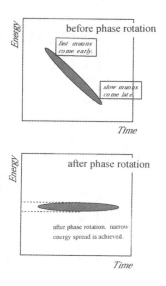

Figure 3. Principle of phase rotation to make the beam energy spread narrower.

synchrotron oscillation that is needed to do phase rotation.

The layout of PRISM based on a FFAG ring is shown in Fig.4. The present design of the FFAG ring has a diameter of about 10 m. The calculations were done to simulate phase rotation in FFAG. It was found that a sinusoidal rf curve is not adequate to carry out complete phase rotation, and rather a saw-tooth rf curve is used instead. The simulation results are shown in Fig.5. From Fig.5, about five turns of muons in the FFAG ring will complete phase rotation.

Since PRISM is focused on experiments with stopped muons, the central muon momentum of the FFAG ring is set to 68 MeV/c (corresponding to a kinetic energy of 20 MeV). From the simulation studies of phase rotation at FFAG, the original momentum spread of ±20 % is reduced down to ±2 % after phase rotation. Its R&D program starts from a relatively low repetition rate (∼ kHz), and aims at a higher repetition in the future. This repetition rate is limited at present by that of a kicker magnet for injection and extraction of a beam to FFAG. The concrete layout of the FFAG machine for PRISM is shown in Fig.6.

PRISM is planned to be constructed at the planned 50-GeV PS at the KEK-JAERI Joint Project [4]. If it is combined with the planned 50-GeV PS

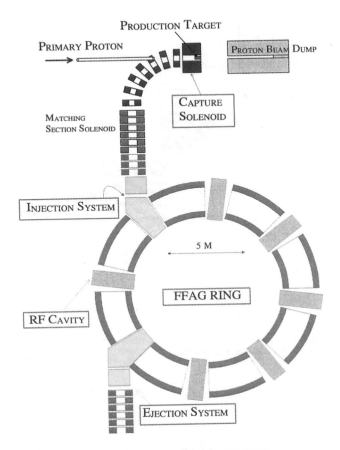

Figure 4. Schematic layout of PRISM.

at the KEK-JAERI Joint Project, PRISM would provide about $10^{12}\mu^{\pm}/\text{sec}$. A possible layout of PRISM in the experimental hall of the 50-GeV PS is shown in Fig.7.[a]

As mentioned before, those ideas of high-field pion capture and phase

---

[a] Since then, instead of using a slowly-extracted proton beam, a fast-extracted proton beam is considered to be used. In this case, protons in the 50-GeV ring can be bunshed in about 100 pulses and then they are extracted in a fast extraction mode as they are in the ring. In this fast extracted proton beam, another dedicated experimental hall for fast-extracted proton beam should be needed.

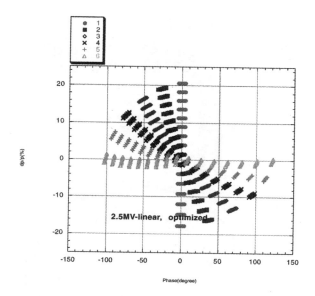

Figure 5. Simulation of phase rotation at FFAG with a saw-tooth RF wave form. After 5 turns, an energy spread becomes ± a few %.

rotation have emerged in studies of a $\mu^+\mu^-$ collider at the high-energy frontier. Although there are many common R&D items between a low-energy muon source and a $\mu^+\mu^-$ collider, there are discussions on whether the front-end muon collider (FMC) could be directly used in experiments with muons. The FMC will run with a pulsed beam of slow repetition (at typically 15 Hz). However, most experiments with muons require a beam with a high duty factor, because of the reduction of the instantaneous rate. Thus, independent R&D items exist in a low-energy muon source, such as in PRISM.

The direct applications of PRISM to the searches for muon LFV process will be covered by the other speaker in this proceedings. Therefore, the application of PRISM to other muon applied-science is presented in the next section.

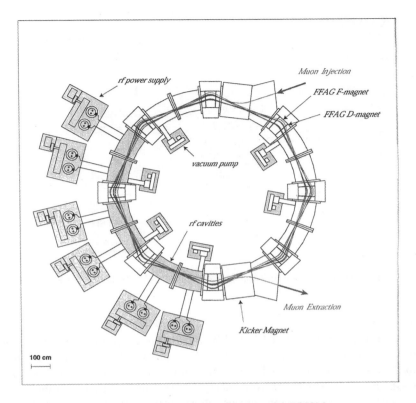

Figure 6. Layout of FFAG machine for PRISM.

5 Application of PRISM

Applications of a muon beam from PRISM to applied-science would be very broad. One of them is a trace-element analysis. Because of a high intensity and narrow energy width, a negative muon beam from PRISM can be effectively stopped and used for very small (valuable) samples such as living cells. For further interesting applications, a brain scanning with muonic X-ray measurement or μ-decay electrons. Here, narrow energy width of the PRISM beam would be very useful to identify a small region of muon stopping. Of course, materials science with muons (μSR) and biological application are already established and high intensity beam would be beneficial. However, higher muon polarization would be necessary.

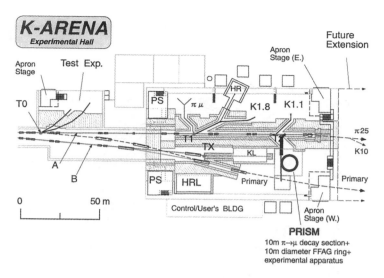

Figure 7. PRISM in the experimental hall of the 50-GeV PS in the KEK/JAERI joint project.

So far, phase rotation at PRISM is considered to produce a narrow energy width starting from a short pulse. The short pulse is very essential for μ^-SR since a rotation frequency is limited by a muon beam width. However, if one starts from a beam with narrow energy width such as a surface muon (μ^+) beam with wide proton beam pulses, phase rotation in the other way around can produce a beam with a narrower timing width. This could be useful to see dynamic behavior, such as nsec response in materials science or biology. Thus, the applications which would be opened for the first time with PRISM are very broad and significant.

6   More on PRISM

The high-intensity muon sources could be used not only for experiments with low-energy muons, but also for experiments with energetic muons if the muons thus produced are injected into additional accelerators for further acceleration. One of the most important physics issue with accelerated muons in PRISM is to measure the electric dipole moment (EDM) of the muon. Here the best muon momentum is around 100 - 500 MeV/c, and the accelerated PRISM beam will be injected into a dedicated ring where the EDM of the muon is measured. This ring could be built with a single steel yoke to maintain an excellent field uniformity. The EDM measurement needs high statistics as well as high polarization and good control of systematic errors.

If a muon from PRISM is accelerated up to about 3 GeV/c, a measurement of $g - 2$ moment of the muon can be considered. The principle idea of possible improvement is to use a beam with a smaller size and reasonable intensity so that integrated field uniformity which the muons in the $g - 2$ ring see would be improved. However, it is important to examine whether other systematic errors can be under control.

The muon ionization cooling in principle can be considered at the PRISM FFAG to reduce the emittance of the muon beam. It can be done with liquid hydrogen absorber, where the muons lose their energy, with an accelerating rf field which restore the longitudinal muon energy. In Monte Carlo simulation, it is promising if a high-gradient rf field is available. Further studies are needed.

7 Summary

A high-intensity low-energy muon beam from PRISM with narrow energy width and high purity would lead a new world of muon science with stopped muons. In particle physics, the muon LFV is one of the most exciting physics issue. There are other interesting applications such as muon-applied science and the measurements of muon magnetic moments and EDM. We hope that the development of PRISM is critical for future success in muon science.

Acknowledgments

The author is grateful to the members of the PRISM working group.

References

1. Y. Kuno and Y. Okada, Review of Modern Physics, **73** 151 (2001).
2. Y. Kuno, *et al.*, *Proceedings of Workshop on High Intensity Secondary Beam with Phase Rotation* edited by Y. Kuno and N. Sasao (1998) p.71.
3. Y. Kuno and Y. Mori, *Proceedings of "High Intensity Muon Sources* (World Scientific) edited by Y. Kuno and T. Yokoi (1999), P.119.
4. The Joint Project Team of JAERI and KEK, "The Joint Project for High-Intensity Proton Accelerators", KEK Report 99-4, JAERI-Tech 99-056, JHF-99-3, (1999).

PRISM
— EXPERIMENT —

M. AOKI

Institute of Particle and Nuclear Studies,
High Energy Accelerator Research Organization,
1-1 Oho, Tsukuba-shi, Ibaraki-ken 305-0801, JAPAN
E-mail: masaharu.aoki@kek.jp

A facility PRISM (**P**hase **R**otated **I**ntense **S**low **M**uon) is currently under an intense design work as one of physics projects for the forthcoming high intensity proton synchrotron in Tokai, Japan[1]. With this high intensity muon beam (10^{12} muons per second), we would be able to push forward physics of lepton flavor violation in near future. The branching ratio of $\mu^- - e^-$ conversion could be studied at a level of 10^{-18}, and muonium-antimuonium conversion could be also studied below a level of $G_{\mathrm{Mu\overline{Mu}}} < 10^{-4} G_F$.

1 Introduction

The Standard Model (SM) has provided very successful description of the Nature. However, since the SM is based on many given assumptions and parameters, it has been believed to be just a low energy approximation of yet unknown super models. Because of this, both theoretical and experimental studies looking for beyond the SM has being greatly attracted[2]. Lepton flavor conservation (LFC) is one of such given assumptions in SM. Presently, LFC is known to be violated at least in the neutral lepton sector as neutrino oscillation[3,4], while no-violation has been experimentally observed in the charged lepton sector.

In recent years, it was discussed that sizable Lepton Flavor Violation (LFV) effects could be expected in supersymmetric (SUSY) grand unified theories (SUSY-GUT)[5,6]. In particular, LFV for muon decayes are predicted to be just a few orders of magnitude below the current experimental sensitivity[7,8,2]. Another models based on SUSY with right-handed Majorana neutrinos also predict a large LFV effect in relation to the observation of neutrino oscillation[9,10,11].

PRISM, which stands for **P**hase **R**otated **I**ntense **S**low **M**uon, is designed to greatly push forward experimental studies of LFV in near future[12]. Among vast varieties of possible experiments at PRISM, I will focus on two experiments, $\mu^- - e^-$ conversion and muonium-antimuonium conversion. I will show their sensitivity estimations in order to demonstrate physics potential of PRISM.

2 $\mu^- - e^-$ Conversion Experiment

When a negative muon is stopped in target materials, it is trapped by an atom and forms a muonic atom. The trapped μ^- cascades down in energy levels in the muonic atom to its $1s$ ground state within a level of ten pico second time range[13]. The fate of the muon is then either decay in an orbit $(\mu^- \to e^- \nu_\mu \overline{\nu_e})$ or capture by a nucleus of mass number A and atomic number Z, $\mu^- + (A, Z) \to \nu_\mu + (A, Z - 1)$, under the consequence of the Standard Model.

In addition to them, there would be the third process which could happen in the context of physics beyond the Standard Model, namely $\mu^- - e^-$ conversion in a muonic atom, $\mu^- + (A, Z) \to e^- + (A, Z)$. This violates the conservation of the lepton flavor numbers, L_μ and L_e by one unit, but conserves the total lepton number, L, thus often called as $|\Delta L_i| = 1$ process.

The strength of this exotic process is usually measured by a branching ratio to the muon capture process,

$$B(\mu^- - e^- \text{ conversion}) = \frac{\Gamma(\mu^- + (A, Z) \to e^- + (A, Z))}{\Gamma(\mu^- + (A, Z) \to \nu_\mu + (A, Z - 1))}, \quad (1)$$

where Γ is the corresponding decay width.

The experimental signature of the $\mu^- - e^-$ conversion is a single mono-energetic electron coming out of the muon stopped material. The peak energy of electron is expressed as $E_{\mu e} \simeq m_\mu - B_\mu$, where m_μ is the muon mass and B_μ is the binding energy of the $1s$ muonic atom. The peak energy depends on the type of material because of different B_μ's. For example, it is 104.3 MeV for titanium and 105.4 MeV for aluminum, respectively. Since none of the potential background processes emit the electrons close to the peak energy except for the very small tail from the muon decay in orbit with $(E_{\mu e} - E_e)^5$ dependence, this process has a potential to improve the sensitivity by using a high muon rate.

The recent experiment at PSI[14] obtained $B(\mu^- - e^- \text{ conversion}) < 6.1 \times 10^{-13}$. A new proposal, MECO[15], will run in several years and will improve $B(\mu^- - e^- \text{ conversion})$ down to a level of 10^{-16}, which covers wide range of the parameter space predicted in SUSY-GUT. Another experiment called PRIME (**PRI**sm **M**uon-**E**lectron) is under intensive design work as a flagship experiment at PRISM

The basic design of the detector for PRIME is based on that of MECO since the MECO detector is quite well optimized. There will be layered thin muon stopping target, electron tracker and electron calorimeter installed in this order from upstream in a graded solenoid magnetic field. The only major difference between PRIME and MECO is a total thickness of the muon stopping target, which makes it possible to achieve an improved sensitivity in PRIME.

2.1 Target Material

In MECO, it has to veto the first 600 nsec after the beam pulse in order to reduce the possible background coming from the beam pion capture. Thus, it has to use aluminum instead of titanium as the muon stopping target since the muon life time in the titanium (329 ns) is too short. On the other hand, the sensitivity to the photonic process in $\mu^- - e^-$ conversion with respect to $\mu \rightarrow e\gamma$ becomes the highest in titanium, and it is about 40% less for aluminum.

In PRIME, the muon beam from PRISM is totally free of pion contamination since beam particles, including pion contamination, should travel through a fixed field alternate gradient (FFAG) ring for almost 150 m. Therefore, it is not necessary to place any veto against the prompt beam timing since no pions are in the beam. It is also possible to use titanium as the target material, and that gives 40% higher sensitivity to the photonic process than aluminum. In overall, PRIME would gain almost factor 3 in physics sensitivity per incoming muon over MECO from these considerations.

2.2 Target Thickness

PRISM will provide muons with very narrow energy spread, and makes it possible to use a much thinner target with even higher muon stopping efficiency than MECO. According to a very simple Monte Carlo calculation, 80% of the muons from PRISM could stop within 20 layers of $50\mu m$ titanium target. The electron energy resolution is estimated to be 350 keV(FWHM). This would give us almost 85% of the electron acceptance with electron detection threshold being $E_e > 103.9$ MeV while the number of background events from muon decay in orbit is estimated to be only less than 0.2 for a single event sensitivity of $R = B(\mu^- - e^-$ conversion) $= 10^{-18}$.

2.3 Sensitivity and Background Estimation

By taking full advantage of the narrow-energy-spread muon beam from PRISM, PRIME is able to achieve $R = 10^{-18}$ or better. This would be almost

Table 1. Comparison between PRIME at PRISM and MECO.

	PRIME	MECO
Muon Intensity	$10^{12}\mu^-/s$	$2 \times 10^{11}\mu^-/s$
Muon Momentum	68 ± 2 MeV/c	$15 \sim 90$ MeV/c
Muon Stopping Ratio	80%	40%
Target Material	Ti ($\tau_\mu = 329$ ns)	Al ($\tau_\mu = 880$ ns)
$\frac{B(\mu^- - e^- \text{ conversion})}{B(\mu \rightarrow e\gamma)}$	$\frac{1}{238}$	$\frac{1}{389}$
Target Thickness	50 μm/disk	200 μm/disk
No. of Target Disks	20	17-25
Electron Momentum Resolution	350 keV(FWHM)	750 keV(FWHM)
Time Window	> 0	> 600 ns
Beam Particles	pure μ^-	μ, π and e
Expected Sensitivity	$< 10^{-18}$	$< 10^{-16}$

two orders of magnitude improvement over MECO. Table 1 summarize the major differences between PRIME and MECO.

There are mainly five types of potential background sources to the signal; muon decay in orbit, radiative muon capture with photon conversion, pion capture with photon conversion, beam electron/muon scattering and cosmic ray. As for the muon decay in orbit, it was already described in the previous section.

The radiative muon capture process, $\mu^- {}^{48}_{22}\text{Ti} \rightarrow \gamma\nu {}^{48}_{21}\text{Sc}$, fakes real event signals when γ converts to electrons via the photon conversion process inside the muon stopping target material. However, the end-point energy of photons from this process is only 89.7 MeV for titanium, while the corresponding energy is very close to the peak energy of $\mu^- - e^-$ conversion for aluminum (102.5 MeV). Thus, different from the case of an aluminum target, it is very unlikely to have electrons with more than 103.9 MeV energy produced from this radiative capture process, unless large errors occur in electron tracking. The number of expected background should be much smaller than that in MECO; could be less than 0.1 at $R = 10^{-18}$.

Pions in the beam are captured by the target atom and produce photons with rather higher energy, typically up to 140 MeV. Thus, electrons from the photon conversion from this process is one of serious background sources in both the past experiments at PSI and incoming experiment, MECO. In contract, it would not be a major background at PRISM since the pion contamination is practically zero (a level of 10^{-17} per muon) due to a long flight path (150 m) of muons in the FFAG ring. However, It could be possible for late pions entering to the FFAG ring right after the main muons finish four turns in the ring. In this case, those late pions only turns less than one turn in the ring. The sizable amount of pions ($\sim 10^{-4}$) would survive and enter the detector. In order to prevent such a case, we could install double kicker magnet at the injection port of the FFAG ring. Because the particle momentum is very low (centered around 68 MeV/c), it is quite easy to sweep out all of the late pions.

Beam electrons with momentum close to the end point momentum might scatter off the muon stopping target and fake the signal electron. Beam muons with momentum more than 77 MeV/c could produce the electron at around the end point energy via Michel decay in flight, and also fake the signal electron. However, in PRISM, beam momentum is only 68 ± 2 MeV/c and thus both processes are below threshold. The expected background would be much less than 0.1 events.

As for the cosmic ray background, it scales the total length of measuring time and the thickness of the muon target. If we simply scale the estimation in MECO, there would be factor 3 reduction due to thinner target in PRIME and factor 1000 reduction due to smaller beam repetition. It would be a level of 0.0001 events.

MECO is a well designed experiment and it is being looked forward to giving the successful result from MECO. After MECO, PRIME would be able to take over it and could push forward the $\mu^- - e^-$ conversion experiment down to a level of $R = 10^{-18}$.

3   Muonium-Antimuonium Conversion Experiment

The spontaneous conversion (or oscillation) of a muonium atom (a hydrogen-like bound state of μ^+ and e^-, $\mu^+ e^-$ or Mu) to its anti-atom (anti-muonium, $\mu^- e^+$ or $\overline{\text{Mu}}$) is also a very interesting class of muon LFV process. In this Mu $-$ $\overline{\text{Mu}}$ conversion, the lepton flavors change by two units ($|\Delta L_i| = 2$) in the ordinary law of separate additive muon and electron numbers, whereas it would be consistent with multiplicative muon or electron number conservation.

There are various models which could induce the Mu $-$ $\overline{\text{Mu}}$ conversion, such as those mediated by doubly charged higgs boson Δ^{++}[16,17], heavy Majorana neutrinos[16], a neutral scaler Φ_N[18] such as a supersymmetric τ-sneutrino $\tilde{\nu}_\tau$[19,20], and a bileptonic flavor diagonal gauge boson X^{++}[21,22]. It was mentioned that a lower limit of $G_{\text{Mu}\overline{\text{Mu}}} \geq (1-40) \times 10^{-4} G_F$ could be even obtained in the context of the first model[17].

The present world-record limit obtained at PSI[23] is $G_{\mathrm{Mu\overline{Mu}}} \leq 3.0 \times 10^{-3}G_F$. In this experiment, high energy e^- from μ^- decay in a $\overline{\mathrm{Mu}}$ atom is measured by cylindrical wire chamber, and low energy e^+ (typically 13.5 eV) was detected by micro-channel plate detectors after electrostatic acceleration to 8 keV. The estimated number of background was more than one, $N_{BG} = 1.7$, and a one (background) event was observed in the signal box.

There are two known major sources of background in this measurement. One is accidental coincidence between an energetic e^- produced by Bhabha scattering of e^+ from μ^+ decay in a muonium and a scattered e^+. The second is the physics background from the $\mu^+ \to e^+\nu_e\overline{\nu_\mu}e^+e^-$ decay (branching ratio being 3.4×10^{-5}). It would not be impossible to reduce those background further, and to improve the $G_{\mathrm{Mu\overline{Mu}}}$ another order of magnitude, by modification of this detector. However, it would be very difficult to improve more than that.

3.1 Activation Method

There was totally different experimental technique developed almost 10 years ago in TRIUMF[24]. In this method, an anti-muonium is detected by having it absorbed in tungsten nucleus, $\mu^- \mathrm{W} \to \nu_\mu\,^{184}\mathrm{Ta}$, at very thin surface of the target material. Tantalium is identified by observing triple coincidence of β decay with 8.7 hours of life time, and successive prompt 414 keV γ decay and delayed γ cascade decays (mostly 921 keV). This method is totally insensitive to the accidental background sources which limitted the PSI experiment. Furthermore, it could be possible to handle almost infinite intensity of the muon beam with this method because there is no active devices.

The only one disadvantage of this method is its small sensitivity. In the TIRUMF measurement, the sensitivity was only $(4.6 \pm 1.1) \times 10^{-7}$ per muon and this is almost four orders of magnitude smaller than that of PSI experiment (3×10^{-3} per muon). Because of this, the counter experiment was the right choice in the past, and the quite impressive improvements has been achived in the last decade.

However, once we come to the physics background limit in the counter experiment, it would be worth to re-consider the activation method. Especially with PRISM, the smallness of the sensitivity would be no longer serious limitation since there will be abundant muons.

3.2 Sensitivity and Background

By closely examinating a detailed breakdown of the sensitivity of the activation method [24], it is believed that it would be possible to improve the detection efficiency by more than one order of magnitude. Firstly, the probability of ^{184}Ta production could be improved by a factor of 3 by using enriched pure ^{184}W material as target instead of natural abandant W in which the ^{184}W abandance is only 30.7%. Secondly, 414 keV γ detection efficiency could be improved by using large a scale (relative efficiency being more than 120%) state-of-the-art Ge detector, more than a factor of 5. Delayed γ detection efficiency could be also improved by a factor of 1.5. As a result, the sensitivity would become a level of 10^{-5} per muon.

Exposing the apparatus to $10^{12}\mu^+$/s from PRISM for 2000 hours with the above detection efficiency would give us more than 3 orders of magnitude of the improvement over the latest PSI result in the $\mathrm{Mu} - \overline{\mathrm{Mu}}$ conversion probability. Thus, $G_{\mathrm{Mu}\overline{\mathrm{Mu}}}$ would be searched below a level of $10^{-4}G_F$.

The potential background sources are beam μ^- contamination and cosmic ray. As for the beam μ^- contamination, it is very essential to reduce it below 3×10^{-14} level. At PRISM with the FFAG ring, it should not be difficult because the FFAG ring naturally select a single charge state. Double stage kicker at the injection sector of FFAG will also block late muons with opposite charge.

As for negative muons from cosmic ray, it would stop uniformly in a bulk of the tangsten target, while an anti-muonium only stops in a very thin surface of the target. Even after taking into account of the recoil of ^{184}Ta, the first layer of only 28 nm depth is affected by anti-muonium exposure. Thus the effect of background from cosmic ray could be directly measured by observing ^{184}Ta contamination in the inner layer.

4 Conclusion

We could expect significant improvement in the experimental studies of LFV for both the $|\Delta L_i| = 1$ and $|\Delta L_i| = 2$ processes at PRISM. The branching ratio of $\mu^- - e^-$ convesion could be studied below a level of 10^{-18}, which is two orders of magnitudes better than the expected MECO sensitivity. The muonium-antimuonium conversion couping constant could be studied below a level of $10^{-4}G_F$.

Acknowledgments

I would like to thank Y. Kuno and A. Olin for valuable conversations. This work was supported in Japan by the Ministry of Education, Science, Sports and Culture under the Grant-in-Aid for Scientific Research on Priority Areas (A)(2), No. 12014216.

References

1. The Joint Project team of JAERI and KEK, "The Joint Project for High-Intensity Proton Accelerators", KEK Report 99-4 (JHF-99-3), 1999.
2. Y.Kuno, and Y. Okada, Rev. Mod. Phys. **73**, 151 (2001).
3. Y. Fukuda, *et al.* (Super-Kamiokande Collaboration), *Phys. Rev. Lett.* **81**, 1158 (1998), and Erratum: 1998, *ibid.* **81** 4279.
4. Y. Fukuda, *et al.* (Super-Kamiokande Collaboration), *Phys. Rev. Lett.* **81**, 1562 (1998).
5. S. Dimopoulos and H. Georgi, *Nucl. Phys.* B **193**, 150 (1981).
6. N. Sakai, *Z. Phys.* C **11**, 153 (1981).
7. R. Barbieri, L.J. Hall, and A. Strumia, *Nucl. Phys.* B **445**, 219 (1995).
8. J. Hisano, T. Moroi, K. Tobe, and M. Yamaguchi *Phys. Lett.* B **391**, 341 (1997), and Erratum: 1997, *ibid*, B **397**, 357.
9. J. Hisano, T. Moroi, K. Tobe, M. Yamaguchi, and T. Yanagida, *Phys. Lett.* B **357**, 576 (1995).
10. J. Hisano, D. Nomura, and T. Yanagida, *Phys. Lett.* B **437**, 351 (1998).
11. J. Hisano, and D. Nomura *Phys. Rev.* D **59**, 116005 (1999).
12. Y. Kuno, *Nucl. Instrum. Methods* A **451**, 233 (2000).
13. A. Czarnecki, *private communication*, 1999.
14. P. Wintz, in *Proceedings of the First International Symposium on Lepton and Baryon Number Violtion*, edited by H.V. Klapdor-Kleingrothaus and I.V. Krivosheina (Institute of Physics Publishing, Bristol and Philadelphia), 534 (1998).
15. M. Bachman *et al.*, (MECO Collaboration), An experimental proposal E940 to Brookhaven National Laboratory AGS, "A Search for $\mu^- N \to e^- N$ with Sensitivity below 10^{-16}", 1997.
16. A. Halprin, *Phys. Rev. Lett.* **48**, 1313 (1982).
17. P. Herczeg and R.N. Mohapatra, *Phys. Rev. Lett.* **69**, 2475 (1992).

18. G.G. Wong and W.S. Hou, *Phys. Rev.* D **50**, R2962 (1994); W.S. Hou and G.G. Wong, *Phys. Rev.* D **53**, 1537 (1996);
19. R.N. Mohapatra, *Z. Phys.* C **56**, 117 (1992).
20. A. Halprin and A. Masiero, *Phys. Rev.* D **48**, 2987 (1993).
21. H. Fujii *et al.*, *Phys. Rev.* D **49**, 559 (1994).
22. P.H. Frampton, *Phys. Rev. Lett.* **69**, 2889 (1992).
23. L. Willmann *et al.*, *Phys. Rev. Lett.* **82**, 49 (1999).
24. T.M. Huber *et al.*, *Phys. Rev.* D **41**, 2709 (1990).

$\mu^+ \to e^+\gamma$ EXPERIMENT AT PSI

J.YASHIMA, T.MASHIMO, S.MIHARA, T.MORI, H.NISHIGUCHI, W.OOTANI,
S.ORITO AND K.OZONE

International Center for Elementary Particle Physics, University of Tokyo, Japan

L.M.BARKOV, A.A.GREBENUK, B.I.KHAZIN AND V.P.SMAKHIN

Budker Institute of Nuclear Physics, Novosibirsk, Russia

J.EGGER, W.D.HEROLD, P.R.KETTLE AND S.RITT

Paul Scherrer Institute, Switzerland

T.HARUYAMA, Y.KUNO, A.MAKI, Y.SUGIMOTO AND A.YAMAMOTO

Institute of Particle and Nuclear Studies, KEK, Tsukuba, Japan

T.DOKE R.SAWADA, S.SUZUKI, K.TERASAWA AND M.YAMASHITA

Advanced Research Institute for Science and Engineering, Waseda University, Tokyo, Japan

The Search for Lepton Flavor Violation(LFV), the new physics search beyond the Standard Model, has a long history since muon was discovered in 1937. A new search for $\mu^+ \to e^+\gamma$ decay is planed with sensitivity of 10^{-14} at Paul Scherrer Institute(PSI),[1] and detectors have been developed. The concept and status of detector development are presented in this talk.

1 Introduction

In the Standard Model, lepton flavor is conserved with varnishing neutrino masses. Introduction of non-zero neutrino masses into the Standard Model also predicts unmeasurably small lepton flavor violation(LFV). On the other hand, many extensions of the Standard Model, such as Supersymmetric Grand Unified Theory(SUSY-GUT), predict LFV at a measurable level. $\mu^+ \to e^+\gamma$ decay, one of the LFV processes, attracts the attention because the prediction of the SUSY-GUT implies the branching ratio as large as just one or two order of magnitude lower than the current upper limit of 1.2×10^{-11} reported by MEGA experiment in 1999.[2] Moreover, the recent discovery of neutrino masses by Super Kamiokande increases the importance of the search for $\mu^+ \to e^+\gamma$ decay. The see-saw mechanism induced by heavy right-handed Gauge singlets is considered to be the most promising candidate to explain the origin of extremely light neutrino masses. The right-handed singlets also induce the off-diagonal slepton mass matrices which contribute to the branching ratio of

$\mu^+ \to e^+ \gamma$.[3]

Thus, a new search for $\mu^+ \to e^+ \gamma$ will provide very important information beyond the Standard Model and is planed with the sensitivity of 10^{-14} at PSI. The experiment will start data taking in 2003.

2 Detector

A schematic view of the detector is shown in Figure 1. Muons with the intensity of 1×10^8 /sec are stopped on the target located at the center of the detector. Arounding the target, "COnstant Bending RAdius(COBRA) spectrometer" is located to measure the momentum, position and arrival time of positrons. COBRA spectrometer consists of a drift chamber system, scintillation timing counter and a thin magnet specially designed to provide a gradient magnetic field. Outside the magnet, we have designed a liquid Xe scintillation detector for γ-ray detection. Scintillation lights emitted from γ-ray are detected by 800 photomultipliers(PMTs) located in liquid Xe.

There are two major background sources to the experiment, the physical background from radiative muon decay, $\mu^+ \to e^+ \nu_e \bar{\nu}_\mu \gamma$, with two neutrinos carrying small energy, and the accidental overlap of a high energy positron close to the edge of Michel decay spectrum and high energy photon(s) from radiative muon decay or annihilation in flight of positrons. The detector performances required to keep background free condition to 10^{-14} level are shown in Table 1. The rate of backgrounds are estimated to be reduced to 10^{-15} level achieving these requirements.

	FWHM
ΔE_e	0.7%
ΔE_γ	1.4%
$\Delta \theta_{e\gamma}$	12 mrad
$\Delta t_{e\gamma}$	150 psec

Table 1. The expected performance

2.1 Beam

The πE5 beam-line at PSI[4] will provide the most intensive continuous muon beam in the world. The continuous beam improves the sensitivity reducing a mount of accidental backgrounds avoiding overlap of muon decay on the target. The beam channel will be set to $\approx$28MeV/c to collect surface muons.

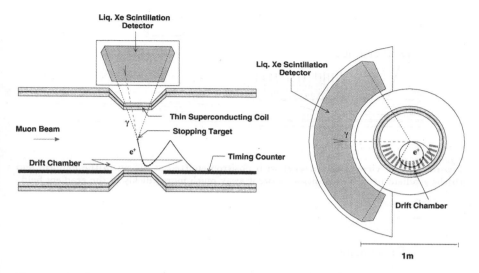

Figure 1. A schematic view of detector, which consists of COBRA spectrometer and liquid Xe photon detector.

The primary protons are provided by the ring cyclotron with the current of 1.8 mA. 10^8 muons/sec are expected to be delivered with a spot size of 5 mm on the target.

2.2 Positron Spectrometer

The major feature of COBRA spectrometer is the gradient magnetic field, which forms the field of approximately 1.1 T in the central region and decreasing as $|z|$ increases. Therefore the positron emitted close to 90°, which makes a lot of turns in the uniform field, is swept away much more quickly as shown in Figure 2. Besides the bending radius of positrons depends only on its momentum independent of its emission angle as shown in Figure 3. This makes it easy to define the momentum window of positron detected by drift chamber.

Magnet
To construct thin wall magnet, we have developed a high-strength aluminum stabilized super conductor, which consists of NbTi multi-filament embedded

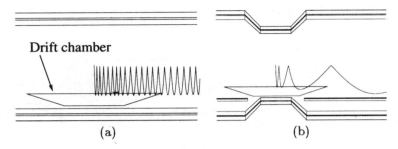

Figure 2. $r - z$ view of the magnet with positron emitted at 88° in (a)the uniform and (b)the gradient field. The positron is swept away much more quickly in the gradient field.

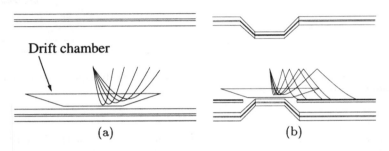

Figure 3. Trajectories of positron with 52.8 MeV emitted various angle in (a)the uniform and (b)the gradient field. The bending radius of positron is independent of the emission angle in the gradient field.

in copper matrix and aluminum stabilizer. The stabilizer is reinforced by "micro-alloying" technology[5] keeping the material as low as possible. Taking into account of cryostat and support structure, the thickness of magnet at central region, where γ-rays pass through before incident into the photon detector, is estimate to be radiation length of 0.15, which is consistent with 95% transparent to a 52.8 MeV γ-ray.

The mechanical analysis is also performed using a finite element analysis method, ANSYS. The maximum stress in the coil is estimated to be less than 180 MPa, which is enough acceptable level. The more detailed design is still under progress and the construction of the magnet will start in 2001.

Drift chamber system
Positrons with energy of 52.8 MeV emitted to $|\cos\theta| < 0.35$ are detected by

the drift chamber system, which consists of 17 sectors of drift chamber aligned at 10° interval. Each sector is composed by two staggered array of drift cells as shown in Figure 4. The wall of chamber is made of thin aluminized kapton foil and aluminum is shaped to make vernier pattern. Charges deposit on these pads are used to determine the position along the sense wire. The chamber is filled by helium based gas to reduce the effect of multiple scattering. The expected position resolutions are $200\mu m$ in r-coordinate and $300\mu m$ in z-coordinate which are consistent with momentum resolution of 0.3 %(σ) and angular resolution of 4 mrad(σ) limited by multiple scattering.

The prototype drift chamber was constructed at PSI to study the engineering problems, gas mixture and its performance. The beam test is performed in the magnetic field of approximately 1T at October 2000. The analysis is now under going.

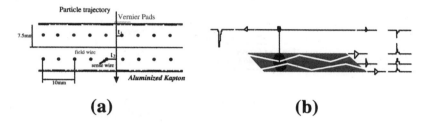

(a)   **(b)**

Figure 4. (a)Cross-sectional view of the drift chamber sector. (b)Charges on the pads induced by a electron avalanche are used to position determination along the wire.

Timing counter

Timing counter for positron detection consists of inner MWPC and outer hodoscope array of plastic scintillators. The MWPC is used to determine the impact point with an accuracy of 1 cm since the impact time and point are very much correlated each other. Scintillators determine the timing with an accuracy of 45 psec(σ).

Two times of beam test were performed at KEK-PS with 0.5~1.0 GeV/c π^+ and positron beam. We used high gain 2-inch PMTs without light-guide to study the intrinsic performance of scintillator at the first beam test. The next beam test was performed to study realistic performance for timing counter with light guides and 1-inch PMTs. The test timing counter consists of 95 cm length scintillator and two PMTs located both ends of scintillator. Figure 5 shows the results of the beam test. The intrinsic resolution of 28 psec

120

for 1 GeV/c π^+ was obtained. The resolutions are proportional to $1/\sqrt{N_{pe}}$ and the ultimate timing resolution is limited by least count of TDC(25psec). These results indicate that the timing resolution of 45 psec(σ) for a 52.8 MeV positron is feasible.

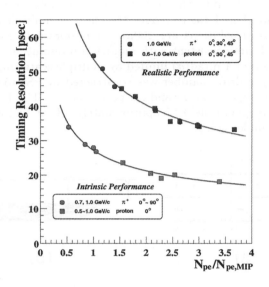

Figure 5. The timing resolution as a function of number of photo-electrons. The number is normalized by minimum ionizing particle($N_{pe,MIP} \sim 1000$)

2.3 Photon detector

Scintillation lights from liquid Xe are detected by 800 PMTs surrounding the effective volume of detector. Liquid Xe scintillator has advantages of high light yield(75% of NaI(TI)), fast signals and short decay times, which are essential ingredients to achieve the requirement for the detector. Liquid Xe is also free from a problem of non-uniformity, which limits the performance of crystal detectors.

In order to be confident on the photon detector, the small prototype was constructed in 1998.[6] The detector has an active area of $116 \times 116 \times 174$ mm^3 viewed by 32 PMTs(HAMAMATSU R6041Q) from all sides. Several γ sources from $0.32 \sim 1.8$ MeV were placed at the end side of box, and α source ^{241}Am was attached at the opposite side for calibration.

Results of energy-, position- and timing-resolution are shown in Figure 6. It is not straightforward to extrapolate these results to the final detector, however simple extrapolations indicate the energy, position and timing resolution of 0.7 %(σ), better then 2mm(σ) and 45 psec(σ) at 52.8 MeV γ-ray.

Because the small prototype is limited to low γ-ray region by its volume, the larger prototype has been developed to demonstrate the performance with high energy γ-ray. The size of the prototype is as large as 1/3 of the final detector and a total of 264 PMTs are located in the liquid Xe. γ-ray of energy up to 40 MeV will be delivered via Compton scattering at the TERAS ring of Electrotechical Laboratory(ETL) in Tsukuba, Japan. The test is planed at the beginning of 2001.

3 Summary

Search for LFV is very important to test physics beyond the Standard Model. We therefore propose to search for $\mu^+ \rightarrow e^+\gamma$ decay with a sensitivity of 10^{-14} branching ratio using continuous muon beam at PSI.

It is essential to reject backgrounds to reach the sensitivity, and we are developing high performance detectors, COBRA spectrometer and liquid Xe photon detector. To start construction of the magnet in 2001, we have developed a high-strength stabilized super conductor. Mechanical analysis is also performed and indicates that the stress in the coil is enough acceptable level. The prototype of drift chamber was constructed at PSI and tested in the magnetic field. The result of beam test will be reported soon. We also performed beam tests for timing counter at KEK-PS in 1998, and results indicate that the resolution of 45 psec is feasible.

A prototype of liquid Xe photon detector was constructed and tested with γ-ray sources. The simple extrapolations to the 52.8 MeV show that the expected performances can be achieved. We are planning to investigate performance to higher energy γ-ray with larger prototype. The test will be performed at the TERAS ring of ETL in the beginning of 2001.

References

1. T.Mori et al., "Search for $\mu^+ \rightarrow e^+\gamma$ down to 10^{-14} branching ratio", Research Proposal to PSI, May 1999.
2. M.L.Brooks et al., Phys.Rev.Lett. 83 (1999) 1521-1524
3. J.Hisano and D.Nomura, Phys.Rev.D59 116005 (1999)
4. A.Badertscher et al., PSI internal report (1989); PSI User's Guide, Accelarater Facilities.

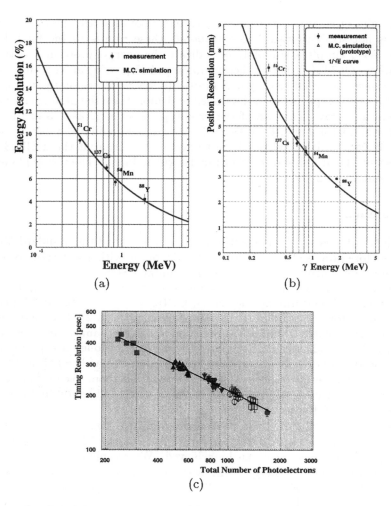

Figure 6. Results of the first prototype. (a)energy resolution, (b)position resolution as a function of energy of γ-ray and (c)timing resolution concerned with number of photoelectrons.

5. A.Yamamoto *et al.*, Nucl.Phys. B78(1999) 565-570
6. K.Abe *et al.*, MEG technical Note TN011(1999)
7. K.Abe *et al.*, MEG technical Note TN012(1999)

LEPTOGENESIS: NEUTRINOS AND NEW LEPTON FLAVOR VIOLATION AT THE TEV ENERGY SCALE

ERNEST MA

Department of Physics, University of California, Riverside,
CA 92521, USA
E-mail: ma@phyun8.ucr.edu

Leptogenesis, i.e. the creation of a lepton asymmetry in the early Universe, may occur through the decay of heavy singlet (right-handed) neutrinos. If we require it not to be erased by physics beyond the Standard Model at the TeV energy scale, then only 2 candidates are possible if they are subgroups of E_6. These 2 solutions happen to be also the only ones within 1σ of the atomic parity violation data and the invisible Z width. Lepton flavor violation is predicted in one model, as well as in another unrelated model of neutrino masses where the observable decay of a doubly charged scalar would determine the relative magnitude of each element of the neutrino mass matrix.

1 Introduction

In the minimal Standard Model, leptons transform under $SU(3)_C \times SU(2)_L \times U(1)_Y$ according to

$$\begin{pmatrix} \nu \\ l \end{pmatrix}_L \sim (1, 2, -\tfrac{1}{2}), \quad l_R \sim (1, 1, -1). \tag{1}$$

In the absence of ν_R, the Majorana mass of the neutrino must come from the effective dimension-5 operator [1]

$$\frac{1}{\Lambda}(\nu_i \phi^0 - l_i \phi^+)(\nu_j \phi^0 - l_j \phi^+). \tag{2}$$

This means that the so-called "seesaw" structure, i.e. $m_\nu \sim v^2/\Lambda$ is inevitable, no matter what specific mechanism is used to obtain m_ν.

The canonical seesaw mechanism [2] is achieved with the addition of a heavy $N_R \sim (1, 1, 0)$. In that case, the interaction $f\bar{N}_R \nu_i \phi^0$ and the large Majorana mass m_N of N_R allow the above effective operator to be realized, with

$$m_\nu = \frac{f^2 \langle \phi^0 \rangle^2}{m_N} = \frac{m_D^2}{m_N}. \tag{3}$$

2 Leptogenesis from N Decay

Consider the decay of N in the early Universe. [3,4] Since it is a heavy Majorana particle, it can decay into both $l^- \phi^+$ (with lepton number $L = 1$) and $l^+ \phi^-$

(with $L = -1$). Hence L is violated. Now CP may also be violated if the one-loop corections are taken into account. Specifically, consider $N_1 \to l^-\phi^+$. This amplitude has contributions from the tree diagram as well as a vertex correction and a self-energy correction, with $l^+\phi^-$ in the intermediate state and $N_{2,3}$ appearing in the cross and direct channels respectively. Calling this amplitude $A + iB$, where A and B are the dispersive and absorptive parts, the asymmetry generated by N_1 decay is then proportional to

$$|A + iB|^2 - |A^* + iB^*|^2 = 4\text{Im}(A^*B), \tag{4}$$

which is nonzero if A and B have a relative phase, i.e. if CP is violated. Note that if there is only one N (i.e. $N_{2,3}$ exchange is absent), then this phase is automatically zero in the above.

In the approximation that M_1 is much smaller than $M_{2,3}$, the decay asymmetry is

$$\delta \simeq \frac{G_F}{2\pi\sqrt{2}} \frac{1}{(m_D^\dagger m_D)_{11}} \sum_{j=2,3} \text{Im}(m_D^\dagger m_D)_{1j}^2 \frac{M_1}{M_j}, \tag{5}$$

which may be washed out by the inverse interactions which also violate L unless the decay occurs out of equilibrium with the rest of the particles in the Universe as it expands. This places a constraint on M_1 to be in the range 10^9 to 10^{13} GeV.

Once the N's have decoupled as the Universe cools, the other (light) particles, i.e. those of the Standard Model, have only L conserving interactions except for the nonperturbative sphalerons which violate $B + L$, but conserves $B - L$. Hence the L asymmetry generated by N decay gets converted [5] into a baryon asymmetry of the Universe, which is observed at present to be of order 10^{-10}.

If N decay is indeed the source of this B asymmetry (to which we owe our own very existence), then any TeV extension of the Standard Model should also conserve $B - L$. In the next section it will be shown that if this extension involves a subgroup of E_6, then there are only 2 possible candidates. [6]

3 Possible E_6 Subgroups at the TeV Scale

Consider the maximal subgroup $SU(3)_C \times SU(3)_L \times SU(3)_R$ of E_6. The fundamental $\underline{27}$ representation is given by

$$\underline{27} = (3,3,1) + (3^*,1,3^*) + (1,3^*,3). \tag{6}$$

The fermions involved are all taken to be left-handed and defined to be

$$(u,d) \sim (3;2,1/6;1,0), \quad h \sim (3;1,-1/3;1,0), \tag{7}$$

$$(d^c, u^c) \sim (3^*; 1, 0; 2, -1/6), \quad h^c \sim (3^*; 1, 0; 1, 1/3), \tag{8}$$

$$(\nu_e, e) \sim (1; 2, -1/6; 1, -1/3), \quad (e^c, N) \sim (1; 1, 1/3; 2, 1/6), \tag{9}$$

$$(E^c, N_E^c), (\nu_E, E) \sim (1; 2, -1/6; 2, 1/6), \quad S \sim (1; 1, 1/3; 1, -1/3), \tag{10}$$

under $SU(3)_C \times SU(2)_L \times U(1)_{Y_L} \times SU(2)_R \times U(1)_{Y_R}$. In this notation, the electric charge is given by $Q = T_{3L} + Y_L + T_{3R} + Y_R$, with $B - L = 2(Y_L + Y_R)$.

Since (e^c, N) is an $SU(2)_R$ doublet, the requirement that $m_N > 10^9$ GeV for successful leptogenesis is not compatible with the existence of $SU(2)_R$ at the TeV scale. This rules out the subgroup $SU(3)_C \times SU(2)_L \times SU(2)_R \times U(1)_{B-L}$ of $SO(10)$. However, as shown below, a different decomposition of $SU(3)_R$, i.e. into $SU(2)'_R \times U(1)_{Y'_R}$, with

$$T'_{3R} = \frac{1}{2}T_{3R} + \frac{3}{2}Y_R, \quad Y'_R = \frac{1}{2}T_{3R} - \frac{1}{2}Y_R, \tag{11}$$

allows N to be trivial under the new skew left-right gauge group[7] so that its existence at the TeV scale is compatible with N leptogenesis.

To see how this works, consider the decomposition of E_6 into its $SO(10)$ and $SU(5)$ subgroups, then

$$\underline{27} = (16, 5^*)[d^c, \nu_e, e] + (16, 10)[u, d, u^c, e^c] + (16, 1)[N]$$
$$+ (10, 5^*)[h^c, \nu_E, E] + (10, 5)[h, E^c, N_E^c] + (1, 1)[S]. \tag{12}$$

If we now switch $(16, 5^*)$ with $(10, 5^*)$ and $(16,1)$ with $(1,1)$, then the $SU(5)$ content remains the same, but the $SO(10)$ does not. The result is a different choice of the direction of $SU(3)_R$ breaking, i.e. V spin instead of the usual T spin. Specifically, we switch d^c with h^c, (ν_e, e) with (ν_E, E), and N with S in Eqs.(8) to (10). Now we may let N be heavy without affecting the new skew left-right gauge group

$$SU(3)_C \times SU(2)_L \times SU(2)'_R \times U(1)_{Y_L + Y'_R}. \tag{13}$$

Note that $B - L$ is conserved by all the interactions of this model at the TeV scale.

Consider next the decomposition $E_6 \to SO(10) \times U(1)_\psi$, then $SO(10) \to SU(5) \times U(1)_\chi$, where

$$Q_\psi = \sqrt{\frac{3}{2}}(Y_L - Y_R), \quad Q_\chi = \sqrt{\frac{1}{10}}(5T_{3R} - 3Y). \tag{14}$$

The arbitrary linear combination $Q_\alpha \equiv Q_\psi \cos\alpha + Q_\chi \sin\alpha$ has been studied extensively as a function of α. If we let $\tan\alpha = 1/\sqrt{15}$, then[8]

$$Q_N = \sqrt{\frac{1}{40}}(6Y_L + T_{3R} - 9Y_R). \tag{15}$$

In that case, N is also trivial under this $U(1)_N$. Hence

$$SU(3)_C \times SU(2)_L \times U(1)_Y \times U(1)_N \qquad (16)$$

is the second and only other possible E_6 extension of the Standard Model compatible with N leptogenesis.

4   New Neutral Currents and Lepton Flavor Violation

In the $SU(3)_C \times SU(2)_L \times SU(2)'_R \times U(1)_{Y_L+Y'_R}$ model, (h^c, u^c) and (e^c, S) are $SU(2)'_R$ doublets, but whereas u^c has $B - L = -1/3$, h^c has $B - L = 2/3$, and whereas e^c has $B - L = 1$, S has $B - L = 0$, hence the W_R^- gauge boson of this model has $B - L = -1$ (because $T'_{3R} = -1$ and $Y'_R = 0$ imply $Y_R = -1/2$). This unusual property has been studied extensively. Moreover, if S is light, it may be considered a "sterile" neutrino. In that case, it has recently been shown [9] that $M_{W_R} > 442$ GeV.

The extra neutral gauge boson Z' of this model is related to W_R by $M_{Z'} = (\cos\theta_W / \cos 2\theta_W)M_{W_R} > 528$ GeV, and it couples to [10]

$$\frac{1}{\sqrt{1-2x}}[xT_{3R} + (1-x)T'_{3R} - xQ]$$

$$= \frac{-1}{\sqrt{1-2x}}[xY_L + \left(\frac{3x-1}{2}\right)T_{3R} - \left(\frac{3-5x}{2}\right)Y_R], \qquad (17)$$

where $x \equiv \sin^2\theta_W$ and $g_L = g_R$. The Z boson of this model behaves in the same way as that of the Standard Model, except

$$Z = \sqrt{1-x}W_L^0 - \frac{x}{\sqrt{1-x}}W_R^0 - \frac{\sqrt{x}\sqrt{1-2x}}{\sqrt{1-x}}B, \qquad (18)$$

which implies a $ZW_R^+W_R^-$ coupling that is absent in the Standard Model.

Together with the Z' of the $U(1)_N$ model, the extra neutral-current interactions of these two E_6 subgroups are the only ones within 1σ of the atomic parity violation data [11] and the invisible Z width. [12] [The $U(1)_N$ model was not considered in Ref. [12], but it can easily be included in their Fig. 1 by noting that it has $\alpha = 0$ and $\tan\beta = \sqrt{15}$ in their notation.] The remarkable convergence of the requirement of successful N leptogenesis and the hint from present neutral-current data regarding possible new physics at the TeV scale is an encouraging sign for the validity of one of these models.

Because of the $ZW_R^+W_R^-$ coupling, lepton flavor violation occurs in one loop through the effective $Z\bar{e}\mu$ vertex. This is the analog of the $ZW_L^+W_L^-$ contribution in the Standard Model. The latter is negligible because all the

μ - e conversion in Al

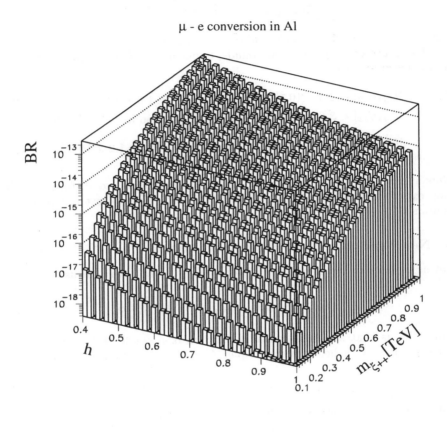

Figure 1: Rate of $\mu - e$ conversion in ^{13}Al.

neutrino masses are very small; the former is not because $m_{S_3} = M_{Z'}$ in the simplest supersymmetric version of this model. [7,9,10] The effective $\mu - e$ transition coupling is then given by

$$g_{Z\bar{e}\mu} = \frac{e^3 U_{\mu 3} U_{e3}}{16\pi^2 \sqrt{x(1-x)}} \left[\frac{r_3}{1-r_3} + \frac{r_3^2 \ln r_3}{(1-r_3)^2} \right], \tag{19}$$

where $r_3 = m_{S_3}^2/M_{W_R}^2 = 1.426$ and $S_{1,2}$ are assumed light.

Using present experimental bounds, upper limits of the mixing of S_3 to μ and e are given below.

$$|U_{\mu 3} U_{e3}| < 2.3 \times 10^{-3} \text{ from } B(\mu \to eee) < 1.0 \times 10^{-12}; \tag{20}$$

$$|U_{\mu 3} U_{e3}| < 3.6 \pm 0.9 \times 10^{-3} \ [\mu - e \text{ conversion in } {}^{48}Ti, \ {}^{208}Pb]; \tag{21}$$

$$|U_{\mu 3} U_{e3}| \frac{M_{W_L}^2}{M_{W_R}^2} < 3.8 \times 10^{-4} \text{ from } B(\mu \to e\gamma) < 1.2 \times 10^{-11}. \tag{22}$$

This shows that unless the mixing angles are extremely small, future precision experiments on lepton flavor violation will be able to test this model in conjunction with TeV colliders.

5 New Verifiable Model of Neutrino Masses

Let us go back to the effective operator of Eq. (2) and rewrite it as

$$\frac{1}{\Lambda}[\nu_i \nu_j \phi^0 \phi^0 - (l_i \nu_j + \nu_i l_j)\phi^0 \phi^+ + l_i l_j \phi^+ \phi^+]. \tag{23}$$

This tells us that another natural realization of a small Majorana neutrino mass is to insert a heavy scalar triplet $\xi = (\xi^{++}, \xi^+, \xi^0)$ with couplings to leptons

$$f_{ij}[\xi^0 \nu_i \nu_j + \xi^+ (\nu_i l_j + l_i \nu_j)/\sqrt{2} + \xi^{++} l_i l_j] + h.c., \tag{24}$$

and to the standard scalar doublet

$$\mu[\bar{\xi}^0 \phi^0 \phi^0 - \sqrt{2}\xi^- \phi^+ \phi^0 + \xi^{--} \phi^+ \phi^+] + h.c. \tag{25}$$

We then obtain [13]

$$m_\nu = \frac{2 f_{ij} \mu \langle \phi^0 \rangle^2}{m_\xi^2} = 2 f_{ij} \langle \xi^0 \rangle. \tag{26}$$

This shows the inevitable seesaw structure, but instead of identifying m_N with the large scale Λ as in the canonical seesaw model [2], we now require only m_ξ^2/μ

to be large. If μ is sufficiently small, the intriguing possibility exists for m_ξ to be of order 1 TeV and be observable at future colliders. The decay

$$\xi^{++} \to l_i^+ l_j^+ \tag{27}$$

is easily detected and its branching fractions **determine** the relative $|f_{ij}|$'s, i.e. the 3×3 neutrino mass matrix up to phases and an overall scale. [14] This possible connection between collider phenomenology and neutrino oscillations is an extremely attractive feature of the proposed Higgs triplet model of neutrino masses.

To understand why μ can be so small and why m_ξ should be of order 1 TeV, one possibility [14] is to consider the Higgs triplet model in the context of large extra dimensions. μ is small here because it violates lepton number and may be represented by the "shining" of a singlet scalar in the bulk, i.e. its vacuum expectation value as felt in our brane. m_ξ is of order 1 TeV because it should be less than the fundamental scale M_* in such theories which is postulated to be of order a few TeV.

Lepton flavor violation in this model may now be predicted if we know f_{ij}. Using a hierarchical neutrino mass matrix which fits present atmospheric [15] and solar [16] neutrino oscillations (choosing the large-angle MSW solution), we predict [14] $\mu - e$ conversion to be easily observable at the MECO experiment as shown in Fig. 1 if m_ξ is indeed of order 1 TeV. The dimensionless parameter h there is proportional to μ.

6 Conclusion

Leptogenesis, neutrino masses, lepton flavor violation, and new physics at future colliders are most likely intertwined. They may well be the different colors of a rainbow (**manoa**) and must exist **together** or not at all.

Acknowledgments

I thank Y. Kuno and W. Molzon for a very useful and stimulating Workshop. This work was supported in part by the U. S. Department of Energy under Grant No. DE-FG03-94ER40837.

References

1. S. Weinberg, Phys. Rev. Lett. **43**, 1566 (1979); E. Ma, Phys. Rev. Lett. **81**, 1171 (1998).
2. T. Yanagida, in *Proceedings of the Workshop on the Unified Theory and the Baryon Number in the Universe*, edited by O. Sawada and A. Sugamoto (KEK Report No. 79-18, Tsukuba, Japan, 1979), p. 95; M. Gell-Mann, P. Ramond, and R. Slansky, in *Supergravity*, edited by P. van Nieuwenhuizen and D. Z. Freedman (North-Holland, Amsterdam, 1979), p. 315; R. N. Mohapatra and G. Senjanovic, Phys. Rev. Lett. **44**, 1316 (1980).
3. M. Fukugita and T. Yanagida, Phys. Lett. **174B**, 45 (1986).
4. M. Flanz, E. A. Paschos, and U. Sarkar, Phys. Lett. **B345**, 248 (1995); Erratum: **B382**, 447 (1996).
5. V. A. Kuzmin, V. A. Rubakov, and M. E. Shaposhnikov, Phys. Lett. **155B**, 36 (1985).
6. T. Hambye, E. Ma, M. Raidal, and U. Sarkar, in preparation.
7. E. Ma, Phys. Rev. **D36**, 274 (1987).
8. E. Ma, Phys. Lett. **B380**, 286 (1996).
9. E. Ma, hep-ph/0002284 (Phys. Rev. **D**, in press).
10. K. S. Babu, X.-G. He, and E. Ma, Phys. Rev. **D36**, 878 (1987).
11. S. C. Bennett and C. E. Wieman, Phys. Rev. Lett. **82**, 2484 (1999).
12. J. Erler and P. Langacker, Phys. Rev. Lett. **84**, 212 (2000).
13. E. Ma and U. Sarkar, Phys. Rev. Lett. **80**, 5716 (1998).
14. E. Ma, M. Raidal, and U. Sarkar, hep-ph/0006046 (Phys. Rev. Lett., in press).
15. H. Sobel, Talk given at *Neutrino 2000*, Sudbury, Canada (June 2000).
16. Y. Suzuki, Talk given at *Neutrino 2000*, Sudbury, Canada (June 2000).

CP VIOLATION AND ATMOSPHERIC NEUTRINOS

DHARAM V. AHLUWALIA

*ISGBG, Escuela de Fisica, Universidad Autonoma de Zacatecas, Ap. Post. C-600,
Zacatecas 98068, Mexico*

YONG LIU

*Institut des Sciences Nucleaires, IN2P3, 53 avenue des Martyrs, 38026 Grenoble
cedex, France*

ION STANCU*

*Department of Physics and Astronomy, Box 870324, University of Alabama,
Tuscaloosa, AL 35487-0324, USA*

We show that if the presently observed L/E-flatness of the electron-like event ratio
in the Super-Kamiokande atmospheric neutrino data is confirmed then the indicated ratio must be *unity*. Further, it is found that once CP is violated the exact
L/E flatness implies: (a) The CP-violating phase, in the standard parameterization, is narrowed down to two possibilities $\pm\pi/2$, and (b) The mixing between the
second and the third generations must be maximal. With these results at hand,
we argue that a dedicated study of the L/E-flatness of the electron-like event ratio
by Super-Kamiokande can serve as an initial investigatory probe of CP violation
in the neutrino sector.

1 Introduction

The Super-Kamiokande data on the atmospheric neutrinos have opened a new
realm of physics research [1]. The simplest interpretation of these data is flavor
oscillations arising from neutrino being linear superposition of some underlying mass eigenstates. This circumstance not only takes us into the physics
beyond the standard model of the high energy physics, but it also allows to
probe various aspects of quantum gravity. As such, much theoretical and experimental effort is being devoted to deciphering the nature of the neutrino.
Here, using a very specific aspect of the Super-Kamiokande data, we shall
analytically constrain the CP-violating neutrino oscillation mixing matrix.
This would help the design of future experiments, allow for more analytically-
oriented theoretical research, and provide a new direction of research at the
existing experimental facilities.

This work joins the on-going research with the observation that as soon as
the first results from the Super-Kamiokande on atmospheric neutrinos became

---

*PRESENTING AUTHOR

available, one of us emphasized that the L/E flatness noted in the *abstract* places a set of constraints on the neutrino oscillation mixing matrix [2]. However, in that, and our subsequent work [3], CP violation has been neglected. Apart from reasons of simplicity, there is no *a priori* reason to assume the absence of CP violation in the neutrino sector. In addition, the observed cosmological baryonic asymmetry may be deeply connected with a CP violation in the leptonic sector [4]. In this work we present a non-trivial generalization of the constraints presented in the early work [2,3] to obtain a CP-violating bimaximal matrix for neutrino oscillations [5].

2   Analytical constraints on the neutrino-oscillation mixing matrix

To generalize the discussion of Refs. [2,3], we start from the probability formula of neutrino oscillations. As in the quark sector, when neutrinos have non-zero masses, their weak eigenstates may not coincide with the mass eigenstates, but may be linear superposition of the mass eigenstates. The latter choice is precisely what is suggested by the existing data [1,6,7,8,9]. As such, in a phenomenology of neutrino oscillations, a flavor eigenstate of a neutrino is postulated to be a linear superposition of some underlying mass eigenstates

$$| \nu_\alpha \rangle = \sum_j U_{\alpha j} | \nu_j \rangle, \tag{1}$$

where $U_{\alpha j}$ is an element of the mixing matrix with $\alpha = e, \mu, \tau$ and $j = 1, 2, 3$ in the framework of three generations. In the literature, U is usually taken as the standard parameterization matrix [10]

$$V = \begin{pmatrix} c_{12}c_{13} & s_{12}c_{13} & s_{13}e^{-i\delta_{13}} \\ -s_{12}c_{23} - c_{12}s_{23}s_{13}e^{i\delta_{13}} & c_{12}c_{23} - s_{12}s_{23}s_{13}e^{i\delta_{13}} & s_{23}c_{13} \\ s_{12}s_{23} - c_{12}c_{23}s_{13}e^{i\delta_{13}} & -c_{12}s_{23} - s_{12}c_{23}s_{13}e^{i\delta_{13}} & c_{23}c_{13} \end{pmatrix} \tag{2}$$

multiplied by a phase matrix

$$P = \begin{pmatrix} 1 & 0 & 0 \\ 0 & e^{i\phi_2} & 0 \\ 0 & 0 & e^{i\phi_3 + \delta_{13}} \end{pmatrix} \tag{3}$$

if neutrinos are of the Majorana type. Here, $c_{ij} = \cos\theta_{ij}$, $s_{ij} = \sin\theta_{ij}$, and ϕ_2 and ϕ_3 are the additional phases for Majorana neutrinos. Due to the unobservable effect of P in flavor oscillation experiments, we will drop it in the discussion and simply equate the mixing matrix U to V in calculations that follow. Furthermore, θ_{12}, θ_{23}, and θ_{13} in U can all be made to lie in the first quadrant by an appropriate re-definition of the relevant fields.

Assuming the underlying mass eigenstates to be relativistic and stable, the flavor-oscillation probability is given by [3,11]

$$P(\nu_\alpha \overset{L}{\to} \nu_\beta) = \delta_{\alpha\beta} - 4\sum_{j<k} \text{Re}(U_{\alpha j}U_{\beta j}^*U_{\alpha k}^*U_{\beta k})\sin^2(\varphi_{jk})$$
$$+ 2\sum_{j<k} \text{Im}(U_{\alpha j}U_{\beta j}^*U_{\alpha k}^*U_{\beta k})\sin(2\varphi_{jk}). \qquad (4)$$

where L (m) refers to the source-detector distance, and the flavor-oscillation inducing kinematic phases φ_{ij}, are defined as

$$\varphi_{ij} = 1.27\,\Delta m_{ij}^2 \frac{L}{E}, \qquad (5)$$

where E (MeV) refers to the "energy" of the flavor state, and $\Delta m_{ij}^2 = m_i^2 - m_j^2$, is the mass-squared difference of the underlying mass eigenstates (eV^2).

For the CP conjugate channel, the CP-odd term, that is, the last term in Eq. (4), changes sign. Note that all $\text{Im}(U_{\alpha j}U_{\beta j}^*U_{\alpha k}^*U_{\beta k})$ with $\alpha \neq \beta$ and $j \neq k$ take the same value $J_{CP} = c_{12}s_{12}c_{23}s_{23}c_{13}^2s_{13}s_\delta$ ($s_\delta = \sin\delta_{13}$, $c_\delta = \cos\delta_{13}$), which is the measure of CP violation [12].

The Super-Kamiokande measured ratio of the e-like events is defined as

$$\mathcal{R}_e = \frac{N_e' + N_{\bar{e}}'}{N_e + N_{\bar{e}}}, \qquad (6)$$

where N_e and $N_{\bar{e}}$ are the numbers of *predicted* ν_e and $\bar{\nu}_e$ events in the absence of neutrino oscillations, whereas the primed quantities are the corresponding numbers of *observed* events, allowing for the presence of neutrino oscillations.

If at the top of atmosphere, i.e. the "source," the number of ν_e ($\bar{\nu}_e$) and ν_μ ($\bar{\nu}_\mu$) are N_{ν_e} ($N_{\bar{\nu}_e}$) and N_{ν_μ} ($N_{\bar{\nu}_\mu}$) respectively, while the cross-sections for ν_e and $\bar{\nu}_e$ are σ_{ν_e} and $\sigma_{\bar{\nu}_e}$; then we obtain the following set of event predictions for the detector:

$$N_e = N_{\nu_e}\sigma_{\nu_e} \qquad (7)$$
$$N_{\bar{e}} = N_{\bar{\nu}_e}\sigma_{\bar{\nu}_e} \qquad (8)$$
$$N_e' = N_{\nu_e}P(\nu_e \overset{L}{\to} \nu_e)\sigma_{\nu_e} + N_{\nu_\mu}P(\nu_\mu \overset{L}{\to} \nu_e)\sigma_{\nu_e} \qquad (9)$$
$$N_{\bar{e}}' = N_{\bar{\nu}_e}P(\bar{\nu}_e \overset{L}{\to} \bar{\nu}_e)\sigma_{\bar{\nu}_e} + N_{\bar{\nu}_\mu}P(\bar{\nu}_\mu \overset{L}{\to} \bar{\nu}_e)\sigma_{\bar{\nu}_e}. \qquad (10)$$

The first two equations correspond to the absence of flavor oscillations, while the last two equations incorporate the effects of flavor oscillations. Now, inserting Eqs. (7-10) into Eq. (6), and taking note of the fact that due to CPT symmetry, $P(\nu_e \overset{L}{\to} \nu_e) = P(\bar{\nu}_e \overset{L}{\to} \bar{\nu}_e)$, we arrive at

$$\mathcal{R}_e - P(\nu_e \overset{L}{\to} \nu_e) = \frac{N_{\nu_\mu} P(\nu_\mu \overset{L}{\to} \nu_e)\sigma_{\nu_e} + N_{\bar{\nu}_\mu} P(\bar{\nu}_\mu \overset{L}{\to} \bar{\nu}_e)\sigma_{\bar{\nu}_e}}{N_{\nu_e}\sigma_{\nu_e} + N_{\bar{\nu}_e}\sigma_{\bar{\nu}_e}}$$

$$= \frac{r}{1 + \lambda x}(P(\nu_\mu \overset{L}{\to} \nu_e) + \lambda y P(\bar{\nu}_\mu \overset{L}{\to} \bar{\nu}_e)), \qquad (11)$$

where we have used the following definitions:

$$\frac{N_{\bar{\nu}_e}}{N_{\nu_e}} = x, \qquad \frac{N_{\bar{\nu}_\mu}}{N_{\nu_\mu}} = y, \qquad \frac{\sigma_{\bar{\nu}_e}}{\sigma_{\nu_e}} = \lambda, \qquad \frac{N_{\nu_\mu}}{N_{\nu_e}} = r. \qquad (12)$$

Now, substituting Eq. (4) into the above equation, and after simplifying, we obtain

$$\left\{ |U_{e1}|^2 |U_{e2}|^2 + r\frac{1+\lambda y}{1+\lambda x}\mathrm{Re}(U_{\mu 1}U_{e1}^* U_{\mu 2}^* U_{e2}) \right\} \sin^2(\varphi_{12})$$

$$+ \left\{ |U_{e1}|^2 |U_{e3}|^2 + r\frac{1+\lambda y}{1+\lambda x}\mathrm{Re}(U_{\mu 1}U_{e1}^* U_{\mu 3}^* U_{e3}) \right\} \sin^2(\varphi_{13})$$

$$+ \left\{ |U_{e2}|^2 |U_{e3}|^2 + r\frac{1+\lambda y}{1+\lambda x}\mathrm{Re}(U_{\mu 2}U_{e2}^* U_{\mu 3}^* U_{e3}) \right\} \sin^2(\varphi_{23})$$

$$- \frac{r}{2}\frac{1-\lambda y}{1+\lambda x}J_{CP}\left[\sin(2\varphi_{12}) + \sin(2\varphi_{13}) + \sin(2\varphi_{23})\right]$$

$$= \frac{1}{4}(1 - \mathcal{R}_e). \qquad (13)$$

It is worth noting that in case $x = y$ and $J_{CP} = 0$, i.e., if the ratio of the numbers of $\bar{\nu}_e$ to ν_e equals the ratio of the numbers of $\bar{\nu}_\mu$ to ν_μ at the source, and if there is no CP violation in the neutrino sector, Eq. (13) looses its dependence on the neutrino and anti-neutrino cross sections.

In order that Eq. (13) holds for all values of L/E we must impose the constraints:

$$\frac{r}{2}\frac{1-\lambda y}{1+\lambda x}J_{CP} = 0 \qquad (14)$$

and

$$|U_{e1}|^2 |U_{e2}|^2 + r\frac{1+\lambda y}{1+\lambda x}\mathrm{Re}(U_{\mu 1}U_{e1}^* U_{\mu 2}^* U_{e2}) = 0 \qquad (15)$$

$$|U_{e1}|^2 |U_{e3}|^2 + r\frac{1+\lambda y}{1+\lambda x}\mathrm{Re}(U_{\mu 1}U_{e1}^* U_{\mu 3}^* U_{e3}) = 0 \qquad (16)$$

$$|U_{e2}|^2 |U_{e3}|^2 + r\frac{1+\lambda y}{1+\lambda x}\mathrm{Re}(U_{\mu 2}U_{e2}^* U_{\mu 3}^* U_{e3}) = 0. \qquad (17)$$

3 The constrained CP-violating matrix

Since Eq. (13) holds for any value of L/E, we are also free to set $L/E = 0$. This yields:

$$\mathcal{R}_e = 1. \tag{18}$$

Although we invoke the Super-Kamiokande observed flatness for $\mathcal{R}_e$ from the beginning, we did *not* refer to a specific value of $\mathcal{R}_e$. The present analysis *predicts* $\mathcal{R}_e$ to be unity. This circumstance is in sharp contrast to the framework of references [2,3] where one assumes both the indicated flatness and the value unity for $\mathcal{R}_e$.

Furthermore, Eq. (14) requires that $J_{CP} = 0$ and/or $\lambda y = 1$. The case in which $J_{CP} = 0$ has been extensively discussed in Ref. [3], and therefore we concentrate here on the case in which $\lambda y = 1$.

According to the definition, $\lambda y = 1$ indicates that if the ratio of the numbers of ν_μ to $\bar{\nu}_\mu$ is close to the ratio of the cross-sections of $\bar{\nu}_e$ to ν_e, then this circumstance allows to ignore the last term on the left hand side of Eq. (13). From Table 1 of Ref. [13] we estimate $y \approx 2.06 \pm 0.31$, [a] while from Ref. [14] we infer $\lambda \approx 1/2.4$. Thus, the required condition is fulfilled on "accidental" grounds. Further justification for ignoring the indicated term lies in the fact that J_{CP} is significantly suppressed by data-indicated $U_{e3} \ll 1$.

Substituting the relevant elements of U into Eqs. (15-17), we obtain

$$c_{12}s_{12}c_{13}^2 + \frac{2r}{1 + \lambda x}\{[c_{12}s_{12}(s_{23}^2 s_{13}^2 - c_{23}^2) + (s_{12}^2 - c_{12}^2)c_{23}s_{23}s_{13}c_\delta] = 0 \tag{19}$$

$$c_{12}s_{13} - \frac{2r}{1 + \lambda x}s_{23}(c_{12}s_{23}s_{13} + s_{12}c_{23}c_\delta) = 0 \tag{20}$$

$$s_{12}s_{13} - \frac{2r}{1 + \lambda x}s_{23}(s_{12}s_{23}s_{13} - c_{12}c_{23}c_\delta) = 0 \tag{21}$$

From Eqs. (20,21) we infer,

$$s_{23}^2 = \frac{1 + \lambda x}{2r} \tag{22}$$

and

$$c_\delta = 0. \tag{23}$$

Thus, the CP phase is $\pi/2$ or $-\pi/2$. Inserting Eqs. (22,23) into Eq. (19), we have

$$c_{23}^2 = \frac{1 + \lambda x}{2r} \tag{24}$$

---

[a] It being the value associated with the lowest atmospheric density in the experiment, identified here as "the top of the atmosphere".

Finally, combining Eq. (22) and Eq. (24), we achieve the results:

$$\theta_{23} = \pi/4, \qquad r = 1 + \lambda x, \tag{25}$$

That is, the mixing between the second and the third generations is maximal, and that the ratio of the numbers of ν_μ to ν_e equals to one plus the ratio of the numbers of $\bar{\nu}_e$ to ν_e events in case of no oscillations.

As a result, the indicated L/E flatness in the the Super-Kamiokande data on the atmospheric neutrinos implies CP-violating maximal mixing matrix:

$$U^\pm = \begin{pmatrix} c_{12}\,c_{13} & s_{12}\,c_{13} & \mp i\,s_{13} \\ -\frac{1}{\sqrt{2}}\,(s_{12} \pm i\,c_{12}\,s_{13}) & \frac{1}{\sqrt{2}}\,(c_{12} \mp i\,s_{12}\,s_{13}) & \frac{1}{\sqrt{2}}c_{13} \\ \frac{1}{\sqrt{2}}\,(s_{12} \mp i\,c_{12}\,s_{13}) & -\frac{1}{\sqrt{2}}\,(c_{12} \pm i\,s_{12}\,s_{13}) & \frac{1}{\sqrt{2}}c_{13} \end{pmatrix} \tag{26}$$

where U^+ corresponds to $\delta_{13} = \pi/2$, and U^- arises from $\delta_{13} = -\pi/2$.

4 Concluding Remarks

Corresponding to the two general forms for U, we obtain the following two measures of CP violation:

$$J_{CP}^\pm = \pm\frac{1}{2}c_{12}s_{12}c_{13}^2 s_{13} = \pm\frac{1}{8}\sin\left(2\theta_{12}\right)\sin\left(2\theta_{13}\right)\cos\left(\theta_{13}\right) \tag{27}$$

In the limit in which θ_{13} vanishes, the $U^\pm$ naturally reduces to the result contained in Eq. (26) of Ref. [3]. Preliminary indications that the U matrix carries a similar form as given in Eq. (26) can also be deciphered from a recent work of Barger, Geer, Raja, and Whisnant [15]. Furthermore, for $\theta_{12} = \pi/4$, U^+ reads

$$U^+ = \begin{pmatrix} c_{13}/\sqrt{2} & c_{13}/\sqrt{2} & -is_{13} \\ -\left(1+is_{13}\right)/2 & \left(1-is_{13}\right)/2 & c_{13}/\sqrt{2} \\ \left(1-is_{13}\right)/2 & -\left(1+is_{13}\right)/2 & c_{13}/\sqrt{2} \end{pmatrix} \tag{28}$$

which coincides with the Xing postulate [16]. The latter, originally invoked to simultaneously allow for the neutrino oscillation explanation of the atmospheric and solar neutrino data, turns out to be dictated upon us by the indicated L/E flatness.

Since the CHOOZ experiment [17] constraints, for large Δm^2, $\sin^2\left(2\theta_{13}\right)$ to be about 0.1, even the large value of $\Delta_{13} = \pm\pi/2$ implied by the present analysis, does not result in a maximal CP-violating difference:

$$P(\nu_\alpha \overset{L}{\to} \nu_\beta) - P(\bar{\nu}_\alpha \overset{L}{\to} \bar{\nu}_\beta) = 4J_{CP}\sum_{j<k}\sin\left(2\varphi_{jk}\right). \tag{29}$$

5 Conclusions

In summary, firstly, our discussion extended in this work seems to obligate us to accept a CP violated neutrino sector. And secondly, once CP is violated in neutrino system, the exact L/E flatness of $\mathcal{R}_e$ implies that: (i) The mixing between the second and the third generations must be maximal, (ii) The ratio $\mathcal{R}_e$ must be unity, (iii) The CP-violating phase in the standard parameterization matrix is $\pi/2$ up to a sign ambiguity, (iv) $N_{\nu_\mu}\sigma_{\nu_e} = N_{\bar{\nu}_\mu}\sigma_{\bar{\nu}_e}$, and finally that (v) $N_{\nu_\mu}/N_{\nu_e} = 1 + N_{\bar{\nu}_e}\sigma_{\bar{\nu}_e}/N_{\nu_e}\sigma_{\nu_e}$.

Therefore, a dedicated study of the ratio $\mathcal{R}_e$ in terms of its precise value, and its L/E dependence, can become a powerful probe to study CP violation in the neutrino sector. Within the framework of this study, if the future data confirms $\mathcal{R}_e$ to be unity for all zenith angles, then we must conclude that either there is no CP violation in the neutrino sector, or it is of the form predicted by Eq. (27). This precise result, in conjunction with knowledge of θ_{12}, θ_{13}, and the associated mass-squared differences, up to a sign ambiguity, completely determines the expectations for CP violation in all neutrino-oscillation channels.

References

1. The Super-Kamiokande collaboration, Y. Fukuda, *et al.*, Phys. Rev. Lett. **81**, 1562 (1998); Phys. Rev. Lett. **82**, 2644 (1999).
2. D. V. Ahluwalia, Mod. Phys. Lett. A **13**, 2249(1998). For other early works on bimaximal mixing, see: V. Barger, S. Pakvasa, T. J. Weiler, K. Whisnant, Phys. Lett. B **437**, 107 (1998); H. Georgi, and S. L. Glashow, Phys. Rev. D **61**, 097301 (2000); A. J. Baltz, A. S. Goldhaber, M. Goldhaber, Phys. Rev. Lett. **81**, 5730 (1998).
3. I. Stancu, D. V. Ahluwalia, Phys. Lett. B **460**, 431 (1999).
4. Y. Liu, and U. Sarkar, Commun. Theor. Phys. (in press); hep-ph/9906307. Also see: H. Fritzsch and Zhi-zhong Xing, Acta Phys. Polon. B **31**, 1349 (2000); K. Fukuura, T. Miura, E. Takasugi, and M. Yoshimura, Phys. Rev. D **61**, 073002 (2000); S. K. Kang, C. S. Kim, and J. D. Kim, hep-ph/0004020; K. Matsuda, N. Takeda, and T. Fukuyama, hep-ph/0003055; G. Barenboim and F. Scheck, Phys. Lett. B **475**, 95 (2000); V. Barger, K. Whisnant, and R. J. N. Phillips, Phys. Rev. Lett. **45**, 2084 (1980); V. Barger, S. Pakvasa, T. J. Weiler, and K. Whisnant, Phys. Lett. B **437**, 107(1998). U. Sarkar and R. Vaidya, Phys. Lett. B **442**, 243 (1998); D. J. Wagner and T. J. Weiler, Phys. Rev. D **59**, 113007 (1999); A. De Rujula, M. B. Gavela, and P. Hernandez, Nucl. Phys. B

547, 21 (1999); A. M. Gago, V. Pleitez, and R. Z. Funchal, Phys. Rev. **61**, 016004 (2000); S. M. Bilenky, C. Giunti, and W. Grimus, Phys. Rev. D **58**, 033001 (1998); M. Tanimoto, Prog. Theor. Phys. **97**, 901 (1997); J. Arafune and J. Sato, Phys. Rev. D **55**, 1653 (1997); T. Fukuyama, K. Matasuda, and H. Nishiura, Phys. Rev. D **57**, 5844 (1998); H. Minakata and H. Nunokawa, Phys. Rev. D **57**, 4403 (1998); J. Arafune, M. Koike, and J. Sato, Phys. Rev. D **56**, 3093 (1997); A. Romanino, Nucl. Phys. B **574**, 675 (2000); G. Barenboim and F. Scheck, Phys. Lett. B **475**, 95 (2000).

5. To avoid confusion, we note in advance that in this paper we distinguish between *flux* and *events*. The former refers to the number of particles of a given species that pass a unit area in a unit time, while the latter depends on the detector sensitivity and the relevant cross sections.

6. Super-Kamiokande collaboration, Y. Fukuda, *et al.*, Phys. Rev. Lett. **81**, 1158 (1998).

7. LSND collaboration, C. Athanassopoulos, *et al.*, Phys. Rev. Lett. **81**, 1774 (1998).

8. K. Eitel, New Jour. Phys. **2**, 1 (2000); KARMEN collaboration, K. Eitel, hep-ex/0008002.

9. T. Ishida, LANL preprint hep-ex/0008047; S. Boyd, LANL preprint hep-ex/0011039.

10. L.-L. Chau and W.-Y. Keung, Phys. Rev. Lett. **53**, 1802 (1984). Particle Data Group, C. Caso *et al.*, Eur. Phys. J. C **3**, 1(1998). V. Barger and K. Whisnant, Phys. Lett. B **456**, 194 (1999). T. Fukuyama, K. Matsuda, and H. Nishiura, Phys. Rev. D **57**, 5844 (1998).

11. K. Dick, M. Freund, M. Lindner, and A. Romanino, Nucl. Phys. B **562**, 29 (1999).

12. C. Jarlskog, Phys. Rev. Lett. **55**, 1039(1985); Z. Phys. C **29**, 491(1985).

13. S. Coutu, *et al.*, Phys. Rev. D **62**, 032001 (2000).

14. G. G. Raffelt, *Stars as laboratories for fundamental physics* (University of Chicago Press, Chicago, 1996), see Eq. (10.17).

15. V. Barger, S. Geer, R. Raja, and K. Whisnant, Phys. Rev. D **62**, 073002 (2000).

16. Zhi-zhong Xing, Phys. Rev. D **61**, 057301 (2000).

17. CHOOZ collaboration, M. Apollonio, *et al.*, Phys. Lett. B **466**, 415 (1999). More recent results from the Palo Verde neutrino oscillation experiment, F. Boehm et al., Phys. Rev. D **62**, 072002 (2000).

18. C.-H. Chang, W.-S. Dai, X.-Q. Li, Y. Liu, F.-C. Ma, Z-J. Tao, Phys. Rev. D **60**, 033006 (1999).

LEPTON FLAVOR VIOLATING ERA OF NEUTRINO PHYSICS

V. BARGER

Physics Department, University of Wisconsin, Madison, WI 53706, USA

The physics agenda for future long-baseline neutrino oscillation experiments is outlined and the prospects for accomplishing those goals at future neutrino facilities are considered. Neutrino factories can deliver better reach in the mixing and mass-squared parameters but conventional super-beams with large water or liquid argon detectors can probe regions of the parameter space that could prove to be interesting.

1 Introduction

Neutrino oscillation phenomena probe the fundamental properties of neutrinos.[1] We presently have evidence for (i) atmospheric ν_μ disappearance oscillations with mass-squared difference $\delta m^2_{\text{atm}} \approx 3 \times 10^{-3} \text{ eV}^2$, (ii) solar ν_e disappearance oscillations with $\delta m^2_{\text{solar}} \approx 5 \times 10^{-5} \text{ eV}^2$, and (iii) and accelerator $\bar{\nu}_\mu \leftrightarrow \bar{\nu}_e$ and $\nu_\mu \leftrightarrow \nu_e$ oscillations with $\delta m^2_{\text{LSND}} \approx 1 \text{ eV}^2$. Limits from accelerator and reactor experiments place important constraints on oscillation possibilities. In particular, reactor experiments exclude large amplitude $\bar{\nu}_e$ disappearance oscillations at $\delta m^2 > 10^{-3} \text{ eV}^2$.

A 3-neutrino model can explain the atmospheric and solar data and provides a useful benchmark for neutrino factory studies. The mixing of 3 neutrinos can be parametrized by 3 angles (θ_{23}, θ_{12}, θ_{13}) and a CP-violation phase (δ). The angle θ_{23} controls the atmospheric oscillation amplitude, θ_{12} controls the solar oscillation amplitude, and θ_{13} couples atmospheric and solar oscillations and controls the amount of ν_e oscillations to ν_μ and ν_τ at the atmospheric scale.

What we now know from experiments is that:

(i) $\theta_{23} \sim \pi/4$, $|\delta m^2_{32}| \sim 3 \times 10^{-3} \text{ eV}^2$ for atmospheric oscillations;

(ii) $\theta_{12} \sim \pi/4$, $|\delta m^2_{32}| \sim 5 \times 10^{-5} \text{ eV}^2$ is favored for solar oscillations (the LAM solution) but other values are not fully excluded;

(iii) $\theta_{13} \sim 0$ ($\sin^2 2\theta_{13} < 0.1$) from the reactor experiments.

In the limit $\theta_{13} = 0$, the oscillations are bimaximal.[2]

A new round of accelerator experiments with medium baselines is under way.[3] The K2K experiment ($L = 250$ km, $\langle E_\nu \rangle \sim 1.4$ GeV) is finding evidence in line with the atmospheric ν_μ disappearance. The MINOS experiment ($L =$

730 km, $\langle E_\nu \rangle \sim 10$ GeV) and the CNGS experiments ICANOE and OPERA ($L = 730$ km, $\langle E_\nu \rangle \sim 20$ GeV) are expected to "see" the first oscillation minimum in $\nu_\mu \leftrightarrow \nu_\mu$, measure $\sin^2 2\theta_{23}$ to 5% and δm^2_{atm} to 10% accuracy, and search for $\nu_\mu \to \nu_e$ down to 1% in amplitude. The short-baseline MiniBooNE experiment at Fermilab will confirm or reject the LSND effect.[4] However, information about neutrino masses and mixing will still be incomplete. Higher intensity beams of both ν_μ and ν_e flavors are needed.

Conventional neutrino beams based on π, K decays are dominantly ν_μ and $\bar{\nu}_\mu$. Neutrino factories would provide high intensity ν_μ and $\bar{\nu}_e$ (or $\bar{\nu}_\mu$ and ν_e) beams from muon decays. The ν_e and $\bar{\nu}_e$ beams would access neutrino oscillation channels that are otherwise inaccessible and are essential for eventual reconstruction of the neutrino mixing matrix.

2 Neutrino Factory

Collimated high-intensity neutrino beams can be obtained from decays of muons stored in an oval ring with straight sections.[5] For $\mu^- \to \nu_\mu e^- \bar{\nu}_e$ decays, the oscillation channels $\nu_\mu \to \nu_\mu$ (disappearance), $\nu_\mu \to \nu_e$ (appearance), and $\nu_\mu \to \nu_\tau$ (appearance) give "right-sign" leptons μ^-, e^-, and τ^-, respectively, whereas the oscillation channels $\bar{\nu}_e \to \bar{\nu}_e$ (disappearance), $\bar{\nu}_e \to \bar{\nu}_\mu$ (appearance), and $\bar{\nu}_e \to \bar{\nu}_\tau$ (appearance) give "wrong-sign" leptons e^+, μ^+, and τ^+, respectively. The oscillation signals are relatively background free. The charge-conjugate channels can be studied in μ^+ decays.

With stored muon energies $E_\mu \gg m_\mu$, the neutrino beam is highly collimated and its flux is $\Phi \simeq N(E_\mu/m_\mu)^2/(\pi L^2)$, where N is the number of useful muon decays and L is the baseline. The νN cross section rises linearly with E_ν and hence with E_μ. The event rate is proportional to $(E_\mu)^3$. The $\bar{\nu} N$ cross section is about $1/2$ of the νN cross section. The charged-current cross section for $\nu_\tau N$ suffers from kinematic suppression at low neutrino energies.

Muons are the easiest to detect. The sign of the muon needs to be measured to distinguish $\nu_\mu \to \nu_\mu$ (right-sign μ) and $\bar{\nu}_e \to \bar{\nu}_\mu$ (wrong-sign μ). The backgrounds to the wrong-sign signal are expected to be small if the energy of the detected μ is $\gtrsim 4$ GeV, which requires stored muon energies $E_\mu \gtrsim 20$ GeV. The sign of electrons is more difficult to determine. It might only be possible to measure the combined $\nu_\mu \to \nu_e$ and $\bar{\nu}_e \to \bar{\nu}_e$ events. Tau-leptons can be detected kinematically or by kinks. The τ-production threshold of $E_\nu = 3.5$ GeV requires higher energy neutrino beams.

The critical parameters of neutrino factory experiments are the number of useful muon decays N, the baseline L, and the data sample size (kt-years), where the latter is defined as the product of the detector fiducial mass, the efficiency of the signal selection requirements, and the number of years of data taking. An entry-level machine may have $N = 6 \times 10^{19}$ and $E_\mu = 20$ GeV, while a high-performance machine may have $N = 6 \times 10^{20}$ and $E_\mu = 35$–70 GeV. The average neutrino beam energies are related to the stored muon energies by $\langle E_{\nu_\mu} \rangle = 0.7 E_\mu$ and $\langle E_{\nu_e} \rangle = 0.6 E_\mu$. Baseline distances from 730 km to 10,000 km are under consideration. For an iron scintillator target a detector mass of 10–50 kt may be employed. With these factories, thousands of neutrino events per year could be realized in a 10 kt detector anywhere on Earth. A number of recent studies have addressed the potential of long-baseline experiments to determine the neutrino masses and mixing parameters.[6–14]

3 Physics Agenda (PA)

There is a well-defined set of physics goals for long-baseline neutrino experiments, as follows.

PA1: The measurement of θ_{13} is a primary goal. A nonzero value of θ_{13} is essential for CP violation and for matter effects with electron-neutrinos. The flavor-changing vacuum probabilities in the leading-oscillation approximation are

$$P(\nu_e \to \nu_\mu) \simeq \sin^2 2\theta_{13} \sin^2 \theta_{23} \sin^2 \left(\frac{\delta m^2_{\text{atm}} L}{4E} \right) , \tag{1}$$

$$P(\nu_e \to \nu_\tau) \simeq \sin^2 2\theta_{13} \cos^2 \theta_{23} \sin^2 \left(\frac{\delta m^2_{\text{atm}} L}{4E} \right) , \tag{2}$$

$$P(\nu_\mu \to \nu_\tau) \simeq \cos^4 \theta_{13} \sin^2 2\theta_{23} \sin^2 \left(\frac{\delta m^2_{\text{atm}} L}{4E} \right) . \tag{3}$$

The $\nu_e \to \nu_\mu$ and $\nu_e \to \nu_\tau$ appearance channels provide good sensitivity to θ_{13}; including the disappearance channels improves the sensitivity. All baselines are okay for a θ_{13} measurement.

PA2: The sign of δm^2_{32} determines the pattern of neutrino masses (i.e., whether the closely spaced mass-eigenstates that give $\delta m^2_{\text{solar}}$ lie above or below the third mass-eigenstate). The coherent scattering of electron neutrinos in matter gives a probability difference $P(\nu_e \to \nu_\mu) - P(\bar{\nu}_e \to \bar{\nu}_\mu)$ that is positive for $\delta m^2_{32} > 0$ and negative for $\delta m^2_{32} < 0$. At baselines of about 2000 km or longer, a proof in principle has been given that the sign of δm^2_{32} can be determined in this way at energies $E_\nu \sim 15$ GeV or higher.[6] A complicating factor is that fake CP violation from matter effects must be distinguished from intrinsic CP violation due to the phase δ.

In the presence of matter the $\nu_e \to \nu_\mu$ probability at small θ_{13} is approximately given by[6]

$$\langle P(\nu_e \to \nu_\mu) \rangle \simeq \frac{\sin^2 2\theta_{13}}{\left|1 - \frac{\langle A \rangle}{\delta m_{32}^2}\right|} \sin^2 \left\{ 1.27 \frac{\delta m_{32}^2 L}{\langle E_\nu \rangle} \left|1 - \frac{\langle A \rangle}{\delta m_{32}^2}\right| \right\}, \qquad (4)$$

where $\langle A \rangle = 2\sqrt{2}\, G_F \langle N_e \rangle \langle E_\nu \rangle$. The sign of A is reversed for $\bar{\nu}_e \to \bar{\nu}_\mu$. Matter effects can enhance appearance rates by an order of magnitude at long baselines. One appearance channel is enhanced and the other suppressed so the separation of the $\nu_e \to \nu_\mu$ and $\bar{\nu}_e \to \bar{\nu}_\mu$ probabilities turns on as L increases.

PA3: Precision measurements of the leading-oscillation parameters at the few percent level are important for testing theoretical models of masses and mixing. The magnitude of δm_{32}^2 affects the shape of the oscillation suppression and $\sin^2 2\theta_{23}$ affects the amount of suppression, so both can be well measured by neutrino factories.

PA4: The subleading $\delta m_{\text{solar}}^2$ oscillation can be probed if the currently favored large-angle mixing (LAM) solution to the solar neutrino problem proves correct. The KAMLAND reactor $\bar{\nu}_e$ experiment should also accurately measure the subleading-oscillation parameters of the LAM solution.[15]

PA5: An important goal of neutrino factories is to detect intrinsic CP violation, $P(\nu_\mu \to \nu_e) \neq P(\bar{\nu}_\mu \to \bar{\nu}_e)$. This is only possible if the solar solution is LAM. Sensitivity to intrinsic CP violation is best for baselines $L = 2000$–4000 km. Intrinsic CP violation at a neutrino factory dominates matter effects for small θ_{13} ($\sim 10^{-4}$), whereas matter effects dominate intrinsic CP for large θ_{13} ($\sim 10^{-1}$).

4 Conventional Neutrino SuperBeams

Conventional neutrino beams are produced from decays of charged pions. These beams of muon-neutrinos have small components of electron-neutrinos. Possible upgrades of existing proton drivers to megawatt (MW) scale are being considered to produce conventional neutrino superbeams.[16] An upgrade to 4 MW of the 0.77 MW beam at the 50 GeV proton synchrotron of the proposed Japan Hadron Facility (JHF) would give an intense neutrino superbeam (SuperJHF) of energy $E_\nu \sim 1$ GeV. An upgrade of the 0.4 MW proton driver at Fermilab would increase the intensity of the NuMI beam by a factor of four (SuperNuMI) with three options for the peak neutrino energy [E_ν(peak) ~ 3 GeV (LE), 7 GeV (ME), and 15 GeV (HE)]. The capabilities of these conventional superbeams to accomplish parts of the neutrino oscillation physics agenda are curently being explored.[16,17] Very large water detectors or smaller liquid argon detectors with excellent background rejection would be used in conjunction with the superbeams.

5 Physics Reach

The results in this section summarize a recent comparative study[17] of super-beam and neutrino-factory physics capabilities in future medium- and long-baseline experiments. The reach of various superbeam and neutrino factory options are compared in Table 1. In these results 3 years of neutrino running followed by 6 years of anti-neutrino running is assumed. For superbeams the argon detector (A) has 30 kt fiducial mass and the water detector (W) a 220 kt fiducial mass, a factor of 10 larger than SuperKamiokande; signal efficiency and estimated detector backgrounds are taken into account. For neutrino factories a 50 kt iron scintillator detector is assumed.

Table 1: Summary of the $\sin^2 2\theta_{13}$ reach (in units of 10^{-3}) for various combinations of neutrino beam, distance, and detector for (i) a 3σ $\nu_\mu \to \nu_e$ appearance with $\delta m_{21}^2 = 10^{-5}$ eV^2, (ii) an unambiguous 3σ determination of the sign of δm_{32}^2 with $\delta m_{21}^2 = 5 \times 10^{-5}$ eV^2, and (iii) a 3σ discovery of CP violation for $\delta m_{21}^2 = 5 \times 10^{-5}$, 1×10^{-4}, and 2×10^{-4} eV^2, from left to right respectively. Dashes in the sign of δm_{32}^2 column indicate that the sign is not always determinable. Dashes in the CPV columns indicate CPV cannot be established for $\sin^2 2\theta_{13} \leq 0.1$, the current experimental upper limit, for any values of the other parameters. The CPV entries are calculated assuming the value of δ that gives the maximal disparity of $N(e^+)$ and $N(e^-)$; for other values of δ, CP violation may not be measurable.

Beam	L (km)	Detector	$\sin^2 2\theta_{13}$ reach (in units of 10^{-3})				
			(i)	(ii)	(iii)		
JHF	295	A	25	—	—	—	25
		W	17	—	—	40	8
SJHF	295	A	8	—	—	5	3
		W	15	—	100	20	5
SNuMI LE	730	A	7	—	100	20	4
		W	30	—	—	—	40
SNuMI ME	2900	A	3	6	—	—	100
		W	8	15	—	—	—
	7300	A	6	6	—	—	—
		W	3	3	—	—	—
SNuMI HE	2900	A	3	7	—	100	20
		W	10	15	—	—	—
	7300	A	4	4	—	—	—
		W	3	3	—	—	—
20 GeV NuF	2900	50 kt	0.5	2.5	—	2	1.5
1.8×10^{20} μ^+	7300		0.5	0.3	—	—	—
20 GeV NuF	2900	50 kt	0.1	1.2	0.6	0.4	0.6
1.8×10^{21} μ^+	7300		0.07	0.1	—	—	—

144

The $\sin^2 2\theta_{13}$ reach at neutrino factories depends on the subleading scale δm_{21}^2. This dependence is illustrated in Fig. 1 for stored muon energies of 20, 30, 40 and 50 GeV.

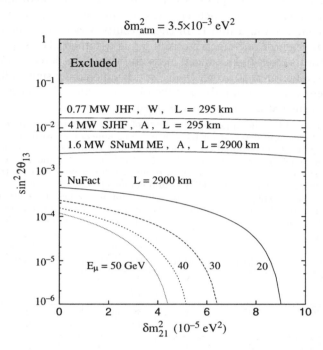

Figure 1: Limiting $\sin^2 2\theta_{13}$ sensitivity for the observation of $\nu_\mu \to \nu_e$ oscillations expected with superbeams and neutrino factories versus the subleading scale δm_{21}^2. (Adapted from the study in Ref. 17.)

The $\text{sign}(\delta m_{32}^2)$ and CP sensitivities are illustrated in Figures 2–5 for various neutrino beam and detector choices:

(i) neutrino factory with $E_\mu = 20$ GeV and $L = 2900$ km (Fig. 2);

(ii) SJHF with a water Cherenkov detector at $L = 295$ km (Fig. 3);

(iii) SNuMI with an $E_\nu(\text{peak}) \sim 3$ GeV beam and a liquid argon detector at $L = 730$ km (Fig. 4);

(iv) SNuMI with an $E_\nu(\text{peak}) \sim 15$ GeV beam and a liquid argon detector at $L = 2900$ km (Fig. 5).

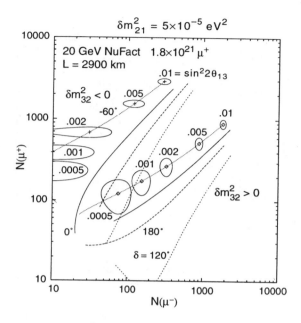

Figure 2: 3σ error ellipses in the $\left[N(\mu^+), N(\mu^-)\right]$-plane, shown for a neutrino factory delivering 3.6×10^{21} useful decays of 20 GeV muons and 1.8×10^{21} useful decays of 20 GeV antimuons, with a 50 kt detector at $L = 2900$ km, for $\delta m_{21}^2 = 5 \times 10^{-5}$ eV^2. The solid and long-dashed curves correspond to the CP-conserving cases $\delta = 0°$ and $180°$, respectively, and the short-dashed and dotted curves correspond to two other cases that give the largest deviation from the CP-conserving curves; along these curves $\sin^2 2\theta_{13}$ varies from 0.0001 to 0.01, as indicated. (From Ref. 17.)

In these SNuMI examples, the CP sensitivity is better in (iii) and the sign(δm_{32}^2) sensitivity is better in (iv).

The conclusions from our comparative study of neutrino factories and conventional superbeams are as follows:

(i) A neutrino factory can deliver between one and two orders of magnitude better reach in $\sin^2 2\theta_{13}$ for $\nu_e \to \nu_\mu$ appearance, the sign of δm_{32}^2, and CP violation. At an $L = 3000$ km baseline there is excellent sensitivity to all three observables. The $\sin^2 2\theta_{13}$ reach is below 10^{-4}. The sign of δm_{32}^2 can be determined and a detection of maximal CP violation made if $\sin^2 2\theta_{13}$ is larger that 10^{-3}.

(ii) Superbeams with a sufficiently ambitious detector can probe $\sin^2 2\theta_{13}$

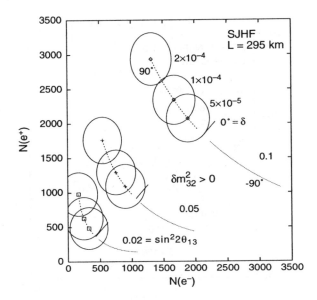

Figure 3: 3σ error ellipses in the $\left[N(e^+), N(e^-)\right]$-plane, shown for the 4 MW SJHF scenario with $L = 295$ km. The contours are for the water Cherenkov detector scenario, with $\sin^2 2\theta_{13} = 0.02$, 0.05, and 0.1. The solid (dashed) [dotted] curves correspond to $\delta = 0°$ $(90°)$ $[-90°]$ with δm^2_{21} varying from 2×10^{-5} eV^2 to 2×10^{-4} eV^2. The error ellipses are shown for three simulated data points at $\delta m^2_{21} = 5 \times 10^{-5}$, 10^{-4} and 2×10^{-4} eV^2. (From Ref. 17.)

down to a few $\times 10^{-3}$. Maximal CP violation may be detected with a JHF or SJHF beam ($E_\nu \sim 1$ GeV) at short baselines, but these facilities will have little sensitivity to sign(δm^2_{32}). Higher-energy superbeams could determine sign(δm^2_{32}) but have little sensitivity to CP violation.

6   Short Baselines

If the LSND effect in $\nu_\mu \rightarrow \nu_e$ and $\bar{\nu}_\mu \rightarrow \bar{\nu}_e$ oscillations is confirmed by Mini-BooNE, then an optimal baseline for future CP-violation studies with a neutrino factory would be

$$L \approx 45 \text{ km} \left(\frac{0.3 \text{ eV}^2}{\delta m^2_{\text{LSND}}}\right) \left(\frac{E_\mu}{20 \text{ GeV}}\right) .$$

(5)

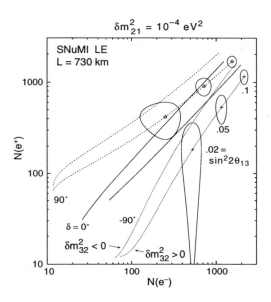

Figure 4: 3σ error ellipses in the $\left[N(e^+), N(e^-)\right]$-plane, shown for the liquid argon detector scenario with the upgraded LE SNuMI beam at $L = 730$ km. The contours are for $\delta m_{21}^2 = 10^{-4}$ eV^2. The solid and long-dashed curves correspond to the CP-conserving cases $\delta = 0°$ and $180°$, respectively, and the short-dashed and dotted curves correspond to two other cases that give the largest deviation from the CP-conserving curves; along these curves $\sin^2 2\theta_{13}$ varies from 0.001 to 0.1, as indicated. (From Ref. 17.)

The distance from Fermilab to Argonne is 30 km, for example. In four-neutrino oscillations, which would be indicated if the LSND, atmospheric, and solar effects are all due to neutrino oscillations, there are 3 CP-violating phases. The size of CP-violating effects in $\nu_e \to \nu_\mu$ and $\nu_\mu \to \nu_\tau$ may be enhanced or reduced relative to three-neutrino oscillations.

7 Overview

We briefly sum up our conclusions regarding future neutrino factory and conventional superbeam studies of neutrino oscillations:

- Three-neutrino mixing and δm^2 parameters can be measured at neutrino factories.

- The amplitude $\sin^2 2\theta_{13}$ is the most crucial parameter. It can be measured down to 10^{-4} at a neutrino factory or to 3×10^{-3} with superbeams.

148

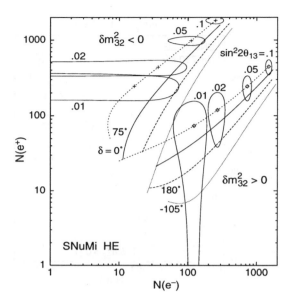

Figure 5: 3σ error ellipses in the $\left[N(e^+), N(e^-)\right]$-plane, shown for the liquid argon detector scenario at $L = 2900$ km with the upgraded HE SNuMI beam. The contours are for $\delta m_{21}^2 = 10^{-4}$ eV^2. The solid and long-dashed curves correspond to the CP-conserving cases $\delta = 0^\circ$ and 180°, respectively, and the short-dashed and dotted curves correspond to two other cases that give the largest deviation from the CP-conserving curves; along these curves $\sin^2 2\theta_{13}$ varies from 0.001 to 0.1, as indicated. (From Ref. 17.)

- A baseline $L \sim 3000$ km is ideal for neutrino factory measurements of $\text{sign}(\delta m_{32}^2)$, CP violation, and the subleading δm_{21}^2 oscillations.

- A longer baseline, $L \sim 7300$ km, is best for precision on δm_{32}^2 and $\sin^2 2\theta_{32}$ at a neutrino factory; for superbeam measurements of $\sin^2 2\theta_{13}$ baselines of 3000–7000 km do equally well.

- Measurements at two baselines would provide complementary advantages (2800 and 7300 for neutrino factories or 295 km and 3000-7000 km for superbeams).

- With the LAM solar solution, intrinsic CP-violating effects could be observable at SuperJHF and at a neutrino factory. The false CP-violation from matter is a serious but manageable complication at long baselines.

Further studies are needed to determine the range of δ for which CP-violating effects are measurable.

- For four-neutrino oscillations, short baselines ($L \sim$ 5–50 km) are also important. With four neutrinos, CP violation occurs at the δm^2_{atm} scale and large effects may be seen.

- Superbeams may be a reasonable next step in exploration of the neutrino sector. However, neutrino factories will eventually be needed for a complete understanding of the neutrino flavor-changing transitions.

Acknowledgments

I thank K. Whisnant for comments and S. Geer, R. Raja, and K. Whisnant for collaboration on this study. This research was supported in part by the U.S. Department of Energy under Grant No. DE-FG02-95ER40896 and in part by the University of Wisconsin Research Committee with funds granted by the Wisconsin Alumni Research Foundation.

1. For recent reviews see e.g., P. Fisher, B. Kayser, and K. McFarland, Ann. Rev. Nucl. Part. Sci. **49**, 481 (1999); V. Barger and K. Whisnant (hep-ph/0006235), J. Learned (hep-ex/0007056), R. Mohapatra (hep-ph/9910365), to be published in *Current Aspects of Neutrino Physics*, ed. by D. Caldwell (Springer-Verlag, Heidelberg, 2001); P. Ramond, hep-ph/0001010, to be published in *Proceedings of the International Workshop on Next Generation Nucleon Decay and Neutrino Detector (NNN 99)*, Stony Brook, NY, Sept. 1999, ed. by M. Diwan (AIP Conf. Proc.).
2. V. Barger, S. Pakvasa, T.J. Weiler, and K. Whisnant, Phys. Lett. **B437** 107 (1998); A.J. Baltz, A.S. Goldhaber, and M. Goldhaber, Phys. Rev. Lett. **81**, 5730 (1998); F. Vissani, hep-ph/9708483.
3. M. Sakuda (K2K collaboration), talk at the XXXth International Conference on High Energy Physics (ICHEP 2000), Osaka, Japan, July 2000; MINOS Collaboration, "Neutrino Oscillation Physics at Fermilab: The NuMI-MINOS Project," NuMI-L-375, May 1998; ICARUS/ICANOE web page at http://pcnometh4.cern.ch/; OPERA web page at http://www.cern.ch/opera/.
4. C. Athanassopoulos et al. (LSND collaboration), Phys. Rev. Lett. **77**, 3082 (1996); **81**, 1774 (1998); G. Mills, talk at *Neutrino-2000*, XIXth International Conference on Neutrino Physics and Astrophysics, Sudbury, Canada, June 2000; A. Bazarko (MiniBooNE collaboration), talk at *Neutrino-2000*.
5. S. Geer, Phys. Rev. **D57**, 6989 (1998), Erratum ibid. **D59**, 039903 (1999).

6. V. Barger, S. Geer, and K. Whisnant, Phys. Rev. **D61**, 053004 (2000); V. Barger, S. Geer, R. Raja, and K. Whisnant, Phys. Rev. **D62**, 013004 (2000); **D62**, 073002 (2000); Phys. Lett. **B485** 379 (2000); hep-ph/0007181, to be published in Phys. Rev. D.

7. H.W. Zaglauer and K.H. Schwarzer, Z. Phys. **C40**, 273 (1988); R.H. Bernstein and S.J. Parke, Phys. Rev. **D44**, 2069 (1991); P. Lipari, Phys. Rev. **D61**, 113004 (2000); A. De Rujula, M.B. Gavela, and P. Hernandez, Nucl. Phys. **B547**, 21 (1999); A. Donini, M.B. Gavela, P. Hernandez, and S. Rigolin, Nucl. Phys. **B574**, 23 (2000); A. Cervera et al., Nucl. Phys. **B579**, 17 (2000).

8. K. Dick, M. Freund, M. Lindner, and A. Romanino, Nucl. Phys. **B562**, 299 (1999); A. Romanino, Nucl. Phys. **B574**, 675 (2000); M. Freund, M. Lindner, S.T. Petcov, and A. Romanino, Nucl. Phys. **B578**, 27 (2000); M. Freund, P. Huber, and M. Lindner, Nucl. Phys. **B585**, 105 (2000); K. Dick, M. Freund, P. Huber, and M. Lindner, Nucl. Phys. **B588**, 101 (2000).

9. Neutrino Factory and Muon Collider Collaboration, D. Ayres et al., physics/9911009; C. Albright et al., hep-ex/0008064.

10. M. Campanelli, A. Bueno, and A. Rubbia, hep-ph/9905240; A. Bueno, M. Campanelli, and A. Rubbia, Nucl. Phys. **B573**, 27 (2000); Nucl. Phys. **B589**, 577 (2000).

11. M. Koike and J. Sato, Phys. Rev. **D61**, 073012 (2000); J. Arafune and J. Sato, Phys. Rev. **D55**, 1653 (1997); M. Koike and J. Sato, hep-ph/9707203; J. Arafune, M. Koike, and J. Sato, Phys. Rev. **D56**, 3093 (1997); T. Ota and J. Sato, hep-ph/0011234.

12. I. Mocioiu and R. Shrock, AIP Conf. Proc. **533**, 74 (2000); Phys. Rev. **D62**, 053017 (2000); J. Pantaleone, Phys. Rev. Lett. **81**, 5060 (1998).

13. P.F. Harrison and W.G. Scott, Phys. Lett. **B476**, 349 (2000); O. Yasuda, hep-ph/0005134; H. Yokomakura, K. Kimura, and A. Takamura, hep-ph/0009141; S.J. Parke and T.J. Weiler, hep-ph/0011247; Z.Z. Xing, Phys. Lett. **B487**, 327 (2000) and hep-ph/0009294; T. Hattori, T. Hasuike, and S. Wakaizumi, Phys. Rev. **D62**, 033006 (2000).

14. JHF LOI, http://www-jhf.kek.jp/

15. V. Barger, D. Marfatia, and D. Wood, hep-ph/0011251; R. Barbieri and A. Strumia, hep-ph/0011307; H. Murayama and A. Pierce, hep-ph/0012075.

16. See e.g., B. Richter, hep-ph/0008222; talks by D. Casper, K. Nakamura, Y. Obayashi, and Y.F. Wang in these proceedings; D. Harris et al., Fermilab study (in progress).

17. V. Barger, S. Geer, R. Raja, and K. Whisnant, hep-ph/0012017.

NEUTRINO OSCILLATIONS WITH FOUR GENERATIONS

OSAMU YASUDA

Department of Physics, Tokyo Metropolitan University
Minami-Osawa, Hachioji, Tokyo 192-0397, Japan
E-mail: yasuda@phys.metro-u.ac.jp

Recent status of neutrino oscillation phenomenology with four neutrinos is reviewed. It is emphasized that the so-called (2+2)-scheme as well as the (3+1)-scheme are still consistent with the recent solar and atmospheric neutrino data.

1 Introduction

There have been several experiments [1,2,3,4,5,6,7,8,9 10,11,12,13] which suggest neutrino oscillations. To explain the solar, atmospheric and LSND data within the framework of neutrino oscillations, it is necessary to have at least four kinds of neutrinos. It has been shown in the two flavor framework that the solar neutrino deficit can be explained by neutrino oscillation with the sets of parameters $(\Delta m_\odot^2, \sin^2 2\theta_\odot) \simeq (\mathcal{O}(10^{-5}\text{eV}^2), \mathcal{O}(10^{-2}))$ (SMA (small mixing angle) MSW solution), $(\mathcal{O}(10^{-5}\text{eV}^2), \mathcal{O}(1))$ (LMA (large mixing angle) MSW solution), $(\mathcal{O}(10^{-7}\text{eV}^2), \mathcal{O}(1))$ (LOW solution) or $(\mathcal{O}(10^{-10}\text{eV}^2), \mathcal{O}(1))$ (VO (vacuum oscillation) solution). At the Neutrino 2000 Conference the Superkamiokande group has updated their data of the solar neutrinos and they reported that the LMA MSW solution gives the best fit to the data [9]. At the same time they also showed that the scenario of pure sterile neutrino oscillations $\nu_e \leftrightarrow \nu_s$ is excluded at 95%CL. It has been known that the atmospheric neutrino anomaly can be accounted for by dominant $\nu_\mu \leftrightarrow \nu_\tau$ oscillations with almost maximal mixing $(\Delta m_{\text{atm}}^2, \sin^2 2\theta_{\text{atm}}) \simeq (10^{-2.5}\text{eV}^2, 1.0)$. Again the Superkamiokande group has announced [10] that the possibility of pure sterile neutrino oscillations $\nu_\mu \leftrightarrow \nu_s$ is excluded at 99%CL. On the other hand, combining the final result of LSND and the negative results by E776 [14] ($\nu_\mu \to \nu_e$), Karmen2 [15] ($\nu_\mu \to \nu_e$) and Bugey [16] ($\bar{\nu}_e \to \bar{\nu}_e$), the oscillation parameter satisfies $0.1 \text{ eV}^2 \lesssim \Delta m^2 \lesssim 8 \text{ eV}^2$ and $8 \times 10^{-4} \lesssim \sin^2 2\theta \lesssim 0.04$ at 99%CL. In this talk I will review the present status of four neutrino scenarios in the light of the recent Superkamiokande data of the solar and atmospheric neutrinos.

2 Mass patterns

In the case of four neutrino schemes there are two distinct types of mass patterns. One is the so-called (2+2)-scheme (Fig. 1(a)) and the other is the

152

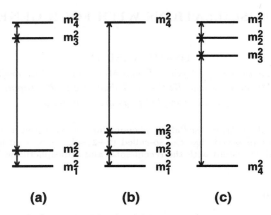

(a) **(b)** **(c)**

Figure 1: Mass patterns of four neutrino schemes. (a) corresponds to (2+2)-scheme, where either ($|\Delta m_{21}^2| = \Delta m_\odot^2$, $|\Delta m_{43}^2| = \Delta m_{atm}^2$) or ($|\Delta m_{43}^2| = \Delta m_\odot^2$, $|\Delta m_{21}^2| = \Delta m_{atm}^2$). (b) and (c) are (3+1)-scheme, where $|\Delta m_{41}^2| = \Delta m_{LSND}^2$ and either ($|\Delta m_{21}^2| = \Delta m_\odot^2$, $|\Delta m_{32}^2| = \Delta m_{atm}^2$) or ($|\Delta m_{32}^2| = \Delta m_\odot^2$, $|\Delta m_{21}^2| = \Delta m_{atm}^2$) is satisfied.

(3+1)-scheme (Fig. 1(b) or (c)). Depending on the type of the two schemes, phenomenology is different.

2.1 (3+1)-scheme

It has been shown in Refs. [18,19] using older data of LSND [12] that the (3+1)-scheme is inconsistent with the Bugey reactor data[16] and the CDHSW disappearance experiment[17] of ν_μ. Let me briefly give this argument in Refs. [18,19]. Without loss of generality I assume that one distinct mass eigenstate is ν_4 (See Fig. 1(b) or (c)) and the largest mass squared difference is $\Delta m_{43}^2 \equiv \Delta m_{LSND}^2$.

In the case of (3+1)-scheme the constraints from the Bugey and CDHSW data are given by

$$1 - P(\bar{\nu}_e \to \bar{\nu}_e) = 4|U_{e4}|^2(1 - |U_{e4}|^2)\Delta_{43} \leq \sin^2 2\theta_{Bugey}(\Delta m_{43}^2)\Delta_{43},$$
$$1 - P(\nu_\mu \to \nu_\mu) = 4|U_{\mu 4}|^2(1 - |U_{\mu 4}|^2)\Delta_{43} \leq \sin^2 2\theta_{CDHSW}(\Delta m_{43}^2)\Delta_{43},$$

respectively, where $\Delta_{43} \equiv \sin^2\left(\Delta m_{43}^2 L/4E\right)$, $\sin^2 2\theta_{Bugey}$ and $\sin^2 2\theta_{CDHSW}$ stand for the values of the boundary of the excluded region in the two flavor analysis as functions of Δm^2 (See Fig. 2). To explain the solar neutrino deficit and the zenith angle dependence of the atmospheric neutrino data it is

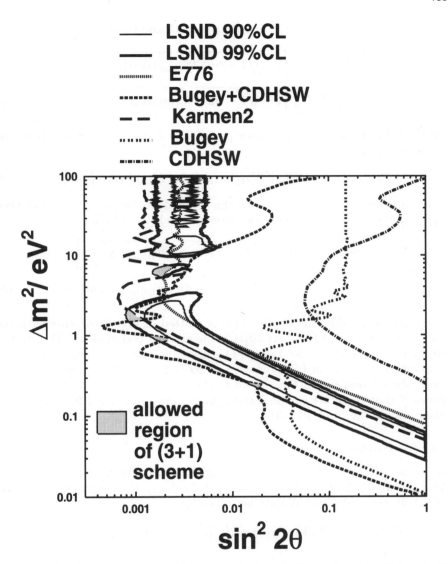

Figure 2: The allowed region of LSND from the final result (the inside of the thick and thin solid lines corresponds to the allowed region at 99%CL and 90%CL, respectively) and the excluded regions of E776, Karmen2, Bugey, CDHSW (the right hand side of each line). The right hand side of the line (Bugey+CDHSW) stands for the excluded region in the case of the (3+1)-scheme. Only the four isolated shadowed areas at $\Delta m^2_{\rm LSND} \simeq 0.3,\ 0.9,\ 1.7,\ 6.0$ eV^2 are consistent with the LSND allowed region at 99%CL in the (3+1)-scheme.

necessary to have $|U_{e4}|^2 < 1/2$ and $|U_{\mu 4}|^2 < 1/2$ and therefore I get

$$|U_{e4}|^2 \leq \frac{1}{2}\left[1 - \sqrt{1 - \sin^2 2\theta_{\text{Bugey}}(\Delta m_{43}^2)}\right]$$

$$|U_{\mu 4}|^2 \leq \frac{1}{2}\left[1 - \sqrt{1 - \sin^2 2\theta_{\text{CDHSW}}(\Delta m_{43}^2)}\right]. \tag{1}$$

On the other hand, the appearance probability $P(\bar{\nu}_\mu \to \bar{\nu}_e)$ of LSND in our scenario is given by

$$P(\bar{\nu}_\mu \to \bar{\nu}_e) = 4|U_{e4}|^2|U_{\mu 4}|^2\Delta_{43} \equiv \sin^2 2\theta_{\text{LSND}}(\Delta m_{43}^2)\Delta_{43}, \tag{2}$$

where $\sin^2 2\theta_{\text{LSND}}(\Delta m_{43}^2)$ stands for the value of $\sin^2 2\theta$ in the LSND allowed region in the two flavor framework. From (1) and (2) I obtain

$$\sin^2 2\theta_{\text{LSND}}(\Delta m_{43}^2) \leq \left[1 - \sqrt{1 - \sin^2 2\theta_{\text{Bugey}}(\Delta m_{43}^2)}\right]$$

$$\times \left[1 - \sqrt{1 - \sin^2 2\theta_{\text{CDHSW}}(\Delta m_{43}^2)}\right]. \tag{3}$$

The value of the right hand side of (3) is plotted in Fig. 2 together with the allowed region of LSND [13]. At 90%CL the allowed region of LSND does not satisfy the condition (3) for the (3+1)-scheme, and actually it used to be the case with older data of LSND [12] even at 99%CL [18,19]. However, in the final result the allowed region has shifted to the lower value of $\sin^2 2\theta$ and it was shown [20] that there are four isolated regions $\Delta m_{\text{LSND}}^2 \simeq 0.3,\ 0.9,\ 1.7,\ 6.0$ eV^2 which satisfy the condition (3).

2.2 (2+2)-scheme

In the case of the (2+2)-scheme, assuming the mass pattern in Fig. 1 (a) with $\Delta m_{21}^2 = \Delta m_\odot^2$, $\Delta m_{32}^2 = \Delta m_{\text{atm}}^2$, $\Delta m_{43}^2 = \Delta m_{\text{LSND}}^2$, the constraints from the LSND, Bugey and CDHSW data are given by

$$P(\bar{\nu}_\mu \to \bar{\nu}_e) = 4|U_{e3}U_{\mu 3}^* + U_{e4}U_{\mu 4}^*|^2\Delta_{32} \equiv \sin^2 2\theta_{\text{LSND}}(\Delta m_{32}^2)\Delta_{32},$$

$$1 - P(\bar{\nu}_e \to \bar{\nu}_e) = 4(|U_{e3}|^2 + |U_{e4}|^2)(1 - |U_{e3}|^2 - |U_{e4}|^2)\Delta_{32}$$
$$\leq \sin^2 2\theta_{\text{Bugey}}(\Delta m_{32}^2)\Delta_{32},$$

$$1 - P(\nu_\mu \to \nu_\mu) = 4(|U_{\mu 3}|^2 + |U_{\mu 4}|^2)(1 - |U_{\mu 3}|^2 - |U_{\mu 4}|^2)\Delta_{32}$$
$$\leq \sin^2 2\theta_{\text{CDHSW}}(\Delta m_{32}^2)\Delta_{32}, \tag{4}$$

where $\Delta_{32} \equiv \sin^2\left(\Delta m_{32}^2 L/4E\right)$.

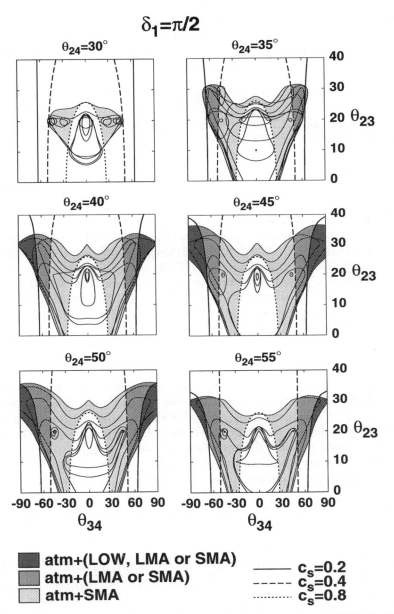

$\delta_1 = \pi/2$

Figure 3: The allowed region of the (2+2)-scheme with $\delta_1 = \pi/2$ at 90 %CL constrained by the Superkamiokande atmospheric neutrino data. The shadowed regions stand for the allowed regions from the combined analysis of the solar and atmospheric neutrino data.

It has been shown [18,19] that these conditions are consistent with all the constraints from the accelerator, reactor data as well as solar and atmospheric neutrino observations. As I will show, to account for both the solar neutrino deficit and the atmospheric neutrino anomaly, it is necessary to have

$$4(|U_{\mu3}|^2 + |U_{\mu4}|^2)(1 - |U_{\mu3}|^2 - |U_{\mu4}|^2) \sim \mathcal{O}(1), \qquad (5)$$

so I take Δm_{32}^2 as small as possible, i.e., $\Delta m_{32}^2 = 0.3$ eV^2 so that (5) be consistent with the CDHSW constraint (4).

3 (2+2)-scheme

3.1 Analysis of the solar neutrino data

The solar neutrino data were analyzed in the framework of the (2+2)-scheme by Ref. [22] on the assumption that $U_{e3} = U_{e4} = 0$, which is justified from the Bugey constraint $|U_{e3}|^2 + |U_{e4}|^2 \ll 1$, and $\Delta m_{\mathrm{atm}}^2$, $\Delta m_{\mathrm{LSND}}^2 \to \infty$ which is also justified since $|\Delta m_{\mathrm{atm}}^2/2E|$, $|\Delta m_{\mathrm{LSND}}^2/2E| \gg \sqrt{2}G_F N_e$ for the solar neutrino problem, where G_F and N_e stand for the Fermi constant and the electron density in the Sun. The conclusion of Ref. [22] is that the SMA MSW solution exists for $0 \le c_s \lesssim 0.8$, while the LMA MSW and LOW solutions survive only for $0 \le c_s \lesssim 0.4$ and $0 \le c_s \lesssim 0.2$, respectively, where $c_s \equiv |U_{s1}|^2 + |U_{s2}|^2$.

3.2 Analysis of the atmospheric neutrino data

The atmospheric neutrino data were analyzed by Refs. [23,24] with the (2+2)-scheme. Refs. [23,24] assumed $U_{e3} = U_{e4} = 0$ as in Ref. [22], and $\Delta m_\odot^2 = 0$ was also assumed. Ref. [23] assumed $\Delta m_{\mathrm{LSND}}^2 = 0.3$ eV^2 so that the result with large $|U_{\mu3}|^2 + |U_{\mu4}|^2$ do not contradict with the CDHSW constraint (4). Ref. [24] did not take into account the contribution from $\Delta m_{\mathrm{LSND}}^2$ to the oscillation probability and their result is a subset of Ref. [23].

Here I adopt the notation in Ref. [18] for the 4×4 MNS matrix:

$$U_{MNS} \equiv R_{34}(\frac{\pi}{2} - \theta_{34})R_{24}(\theta_{24})R_{23}(\frac{\pi}{2})U_{23}(\theta_{23}, \delta_1)U_{14}(\theta_{14}, \delta_3)U_{13}(\theta_{13}, \delta_2)R_{12}(\theta_{12})$$

$$(6)$$

where $U_{23}(\theta_{23}, \delta_1) \equiv e^{2i\delta_1\lambda_3}R_{23}(-\theta_{23})e^{-2i\delta_1\lambda_3}$, $U_{14}(\theta_{14}, \delta_3) \equiv e^{\sqrt{6}i\delta_3\lambda_{15}/2}R_{14}(\theta_{14})$ $e^{-\sqrt{6}i\delta_3\lambda_{15}/2}$, $U_{13}(\theta_{13}, \delta_2) \equiv e^{2i\delta_2\lambda_8/\sqrt{3}}R_{13}(\theta_{13})e^{-2i\delta_2\lambda_8/\sqrt{3}}$, $R_{jk}(\theta) \equiv \exp{(iT_{jk}\theta)}$, $(T_{jk})_{\ell m} = i(\delta_{j\ell}\delta_{km} - \delta_{jm}\delta_{k\ell})$, $2\lambda_3 = \mathrm{diag}(1, -1, 0, 0)$, $2\sqrt{3}\lambda_8 \equiv \mathrm{diag}(1, 1, -2, 0)$, $2\sqrt{6}\lambda_{15} \equiv \mathrm{diag}(1, 1, 1, -3)$ are 4×4 matrices (λ_j are elements of the $su(4)$ generators). With the assumptions $\Delta m_{21}^2 = 0$, $|U_{e4}|^2 = s_{14}^2 = 0$, $|U_{e3}|^2 = c_{14}^2 s_{13}^2 =$

0, θ_{12}, θ_{13} and θ_{14} disappear from U and ν_e decouples from other three neutrinos. Thus the problem is reduced to the three flavor neutrino analysis among ν_μ, ν_τ, ν_s and the reduced MNS matrix is

$$\tilde{U} \equiv \begin{pmatrix} U_{\mu 2} & U_{\mu 3} & U_{\mu 4} \\ U_{\tau 2} & U_{\tau 3} & U_{\tau 4} \\ U_{s2} & U_{s3} & U_{s4} \end{pmatrix} = e^{i(\frac{\pi}{2}-\theta_{34})\lambda_7} D^{-1} e^{i\theta_{24}\lambda_5} D \ e^{i(\theta_{23}-\frac{\pi}{2})\lambda_2},$$

with $D \equiv \mathrm{diag}\left(e^{i\delta_1/2}, 1, e^{-i\delta_1/2}\right)$ (λ_j are the 3×3 Gell-Mann matrices) is the reduced 3×3 MNS matrix. This MNS matrix $\tilde{U}$ is obtained by substitution $\theta_{12} \to \theta_{23} - \pi/2$, $\theta_{13} \to \theta_{24}$, $\theta_{12} \to \pi/2 - \theta_{34}$, $\delta \to \delta_1$ in the standard parametrization in Ref. [27]. It turns out that θ_{34} corresponds to the mixing of $\nu_\mu \leftrightarrow \nu_\tau$ and $\nu_\mu \leftrightarrow \nu_s$, while θ_{23} is the mixing of the contribution of $\sin^2\left(\Delta m_{\mathrm{atm}}^2 L/4E\right)$ and $\sin^2\left(\Delta m_{\mathrm{LSND}}^2 L/4E\right)$ in the oscillation probability. The allowed region at 90%CL of the atmospheric neutrino data is given by the area bounded by thin solid lines in Fig. 3 for $\delta_1 = \pi/2$. The allowed regions for $\delta_1 = 0, \pi/4$ are given in Ref. [23].

3.3 Combined analysis of the solar and atmospheric neutrino data

In Fig. 3, the lines given by $c_s \equiv |U_{s1}|^2 + |U_{s2}|^2 = |c_{23}c_{34} + s_{23}s_{34}s_{24}e^{i\delta_1}|^2 = 0.2, 0.4, 0.8$ are depicted together with the allowed region of the atmospheric neutrino data. By combining the analyses of Ref. [22] and Ref. [23], I obtain the region which satisfies the constraints of the solar and atmospheric neutrino data. The darkest, medium and lightest shadowed areas stand for $\nu_{\mathrm{atm}} + \nu_\odot$(SMA, LMA or LOW), $\nu_{\mathrm{atm}} + \nu_\odot$(SMA or LMA), $\nu_{\mathrm{atm}} + \nu_\odot$(SMA), respectively. Although this result is not quantitative, it gives us a sense on how likely the (2+2)-scheme is allowed by combining the solar and atmospheric neutrino data. Let me emphasize that non-zero contribution of $\sin^2\left(\Delta m_{\mathrm{LSND}}^2 L/4E\right)$ (i.e., the case of $\theta_{23} > 0$) to the oscillation probability is important particularly for the LMA and LOW solar solutions. The region of $\theta_{23} > 0$ has not been analyzed by Ref. [24]. Let me also stress that both the solar neutrinos and the atmospheric neutrinos are accounted for by hybrid of active and sterile oscillations in the (2+2)-scheme.

4 (3+1)-scheme

After the work of Barger et al. [20], people [25,26] have investigated various consequences of the (3+1)-scheme. Here let me make two comments on the (3+1)-scheme.

158

4.1 Atmospheric neutrinos

As in the case of the (2+2)-scheme, I assume $U_{e3} = U_{e4} = \Delta m_\odot^2 = 0$ for simplicity. Then ν_e once again decouples from ν_e, ν_μ, ν_τ and the probability in vacuum for the atmospheric neutrino scale is given by

$$P(\nu_\mu \to \nu_\mu) = 4|U_{\mu3}|^2(1 - |U_{\mu3}|^2 - |U_{\mu4}|^2)\Delta_{32} + 2|U_{\mu4}|^2(1 - |U_{\mu4}|^2),$$
$$P(\nu_\mu \to \nu_\tau) = 4\Re\left[U_{\mu3}U_{\tau3}^*(U_{\mu3}^*U_{\tau3} + U_{\mu4}^*U_{\tau4})\right]\Delta_{32} + 2|U_{\mu4}|^2|U_{\tau4}|^2,$$
$$P(\nu_\mu \to \nu_s) = 4\Re\left[U_{\mu3}U_{s3}^*(U_{\mu3}^*U_{s3} + U_{\mu4}^*U_{s4})\right]\Delta_{32} + 2|U_{\mu4}|^2|U_{s4}|^2,$$

$$(7)$$

where I have taken $\Delta m_{32}^2 \equiv \Delta m_{\rm atm}^2$, $\Delta m_{43}^2 \equiv \Delta m_{\rm LSND}^2$ and I have averaged over rapid oscillations: $\sin^2\left(\Delta m_{\rm LSND}^2 L/4E\right) \to 1/2$. Since the (3+1)-scheme is allowed only for four discrete values of $\Delta m_{\rm LSND}^2$, let me discuss $\Delta m_{\rm LSND}^2 = 0.3$ eV^2 ($|U_{\mu4}|^2 \gtrsim 0.34$) and $\Delta m_{\rm LSND}^2 = 0.9$ eV^2 ($|U_{\mu4}|^2 \simeq 0.03$), 1.7 eV^2 ($|U_{\mu4}|^2 \simeq 0.01$), 6.0 eV^2 ($|U_{\mu4}|^2 \simeq 0.02$), separately. For simplicity I assume $\delta_1 = 0$ since the existence of the CP phase δ_1 does not change the situation very much.

4.1.1 $\Delta m_{\rm LSND}^2 = 0.3$ eV^2

Since we know from the Superkamiokande atmospheric neutrino data that the coefficient of $\sin^2\left(\Delta m_{\rm atm}^2 L/4E\right)$ in $P(\nu_\mu \to \nu_\mu)$ has to be large to have a good fit, I optimize $4|U_{\mu3}|^2(1 - |U_{\mu3}|^2 - |U_{\mu4}|^2)$ with respect to θ_{23} for $|U_{\mu4}|^2 = 0.34$. When $U_{e4} = 0$ I have $|U_{\mu3}|^2 = c_{23}^2 c_{24}^2$, $|U_{\mu4}|^2 = s_{24}^2$, $|U_{\tau4}|^2 = c_{24}^2 c_{34}^2$, $|U_{s4}|^2 = c_{24}^2 s_{34}^2$ in the notation of Ref.[18], and it is easy to see

$$4|U_{\mu3}|^2(1 - |U_{\mu3}|^2 - |U_{\mu4}|^2) = c_{24}^4 \sin^2 2\theta_{23} \le c_{24}^4 = 0.44,$$

where equality holds when $\theta_{23} = \pi/4$. This is the value of θ_{23} for which the fit of the (3+1)-scheme to the atmospheric neutrino data is supposed to be the best for $\Delta m_{\rm LSND}^2 = 0.3$ eV^2. When $\theta_{23} = \pi/4$ the probability in vacuum becomes

$$P(\nu_\mu \to \nu_\tau) = \left(c_{24}^2 s_{34}^2 - \frac{1}{4}\sin^2 2\theta_{24}c_{34}^2\right)\Delta_{32} + \frac{1}{2}c_{34}^2\sin^2 2\theta_{24}$$

$$P(\nu_\mu \to \nu_s) = \left(c_{24}^2 c_{34}^2 - \frac{1}{4}\sin^2 2\theta_{24}s_{34}^2\right)\Delta_{32} + \frac{1}{2}s_{34}^2\sin^2 2\theta_{24}.$$

$$(8)$$

From (8) θ_{34} turns out to be the mixing of $\nu_\mu \leftrightarrow \nu_\tau$ and $\nu_\mu \leftrightarrow \nu_s$ as in the (2+2)-scheme.

I found from the explicit numerical calculation [28] that the fit of the $(3+1)$-scheme with $\Delta m^2_{\text{LSND}}=0.3$ eV^2, $|U_{\mu 4}|^2 = 0.34$, $\theta_{23} = \pi/4$ to the atmospheric neutrino data is very bad for any value of θ_{34} and the region of $\Delta m^2_{\text{LSND}}=0.3$ eV^2 is excluded at 6.9σCL.

4.1.2 Δm^2_{LSND} =0.9, 1.7, 6.0 eV^2

In this case $|U_{\mu 4}|^2 \lesssim 0.03$ and I can put $U_{\mu 4} = 0$ as a good approximation. Then the constant part in the oscillation probability disappears and this case is reduced to the analysis in the (2+2)-scheme with $\theta_{23} = 0$. The allowed region at 90%CL is given roughly by $-\pi/4 \lesssim \theta_{34} \lesssim \pi/4$, $0.8 \lesssim \sin^2 2\theta_{24} \le 1$, where θ_{34} and θ_{24} stand for the mixing of $\nu_\mu \leftrightarrow \nu_\tau$ and $\nu_\mu \leftrightarrow \nu_s$ and the mixing of atmospheric neutrino oscillations, respectively.

4.2   Oscillations of high energy neutrinos in matter

When $|U_{e4}|^2$, $|U_{\mu 4}|^2$ and $|U_{\tau 4}|^2$ are all small, it is naively difficult to distinguish the $(3+1)$-scheme from the ordinary three flavor scenario. However, because of the existence of the small mixing angles in U_{e4}, $U_{\mu 4}$ and the large mass squared difference Δm^2_{LSND} the oscillation probability in matter can have enhancement which never happens in the three flavor case. By taking $\theta_{12} = \pi/4$, $\theta_{13} = 0$, $\theta_{23} = \pi/4$, $\theta_{14} = \epsilon$ ($|\epsilon| \ll 1$), $\theta_{24} = \delta$ ($|\delta| \ll 1$), $\theta_{34} = \pi/2$ in (6) I get

$$
U \simeq \begin{pmatrix} \frac{1}{\sqrt{2}} & \frac{1}{\sqrt{2}} & 0 & \epsilon \\ \frac{1}{2} & -\frac{1}{\sqrt{2}} & \frac{1}{\sqrt{2}} & \delta \\ -\frac{1}{2} & \frac{1}{2} & \frac{1}{\sqrt{2}} & 0 \\ -\frac{\epsilon}{\sqrt{2}} - \frac{\delta}{2} & -\frac{\epsilon}{\sqrt{2}} + \frac{\delta}{2} & 0 & 1 \end{pmatrix} \begin{pmatrix} \nu_1 \\ \nu_2 \\ \nu_3 \\ \nu_4 \end{pmatrix}
$$

which is the same as the MNS matrix in Ref. [20] up to the phase of each factor. The probability $P(\nu_\mu \to \nu_\mu)$ turns out to receive significant deviation from the vacuum one due to the matter effect for $E_\nu \sim \mathcal{O}(1)$ TeV, and the behaviors of $1 - P(\nu_\mu \to \nu_\mu)$ are shown in Fig. 4, where three cases of $|\Delta m^2_{\text{LSND}}|$=0.9 eV^2 ($8.8° \le \epsilon \le 12.2°$, $6.4° \le \delta \le 8.9°$), 1.7 eV^2 ($7.5° \le \epsilon \le 10.2°$, $5.6° \le \delta \le 7.7°$), 6.0 eV^2 ($7.5° \le \epsilon \le 7.7°$, $10.0° \le \delta \le 10.2°$) are considered.[a] The appearance channel which is enhanced is dominantly $\nu_\mu \to \nu_s$, so it may be difficult to detect signs of this enhancement from observations of high energy

---

[a] The eigenvalues of the mass matrix in this case turn out to be roots of a cubic equation and analytic treatment of the oscillation probability is difficult, unlike the cases of three flavors [29] or four flavors [30,25], where one mass scale is dominant and the eigenvalues are roots of a quadratic equation.

160

cosmic neutrinos of energy $E_\nu \sim \mathcal{O}(1)$ TeV, although this enhancement may be observed through neutral current interactions in the future.[b]

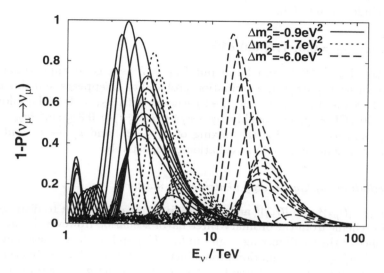

Figure 4: Enhancement of neutrino oscillations due to matter effect in the case of the (3+1)-scheme, where $\theta_{23} = 45°$ is taken for all the cases. For each value of Δm^2= -0.9 eV^2 ($\epsilon = 10.0°$, $\delta = 7.5°$), -1.7 eV^2 ($\epsilon = 5.6°$, $\delta = 7.5°$), -6.0 eV^2 ($\epsilon = 6.4°$, $\delta = 8.8°$), ten curves correspond to $\cos\Theta = -1.0, -0.9, \cdots, -0.1$ from the left to the right, where the zenith angle Θ is related to the baseline L by $L = -2R\cos\Theta$ with R=6378km. Most of the channel is $\nu_\mu \to \nu_s$.

5 Big Bang Nucleosynthesis

It has been shown[32] in the two flavor framework that if sterile neutrino have oscillations with active ones and if $\Delta m^2 \sin^4 2\theta \gtrsim 3 \times 10^{-3}$ eV^2 is satisfied then sterile neutrinos would have been in thermal equilibrium and the number N_ν of light neutrinos in Big Bang Nucleosynthesis (BBN) would be 4. This argument was generalized to the four neutrino case[18,21] and by imposing all the constraints from accelerators, reactors, solar neutrinos and atmospheric neutrinos as well as the BBN constraint $N_\nu < 4.0$ it was concluded that the only consistent four neutrino scenario is the (2+2)-scheme with the MNS mixing

---

[b] Similar enhancement has been discussed in a different context by Ref.[31]. I thank Athar Husain for bringing my attention to Ref.[31].

matrix

$$
U_{MNS} = \begin{pmatrix} U_{e1} & U_{e2} & U_{e3} & U_{e4} \\ U_{\mu1} & U_{\mu2} & U_{\mu3} & U_{\mu4} \\ U_{\tau1} & U_{\tau2} & U_{\tau3} & U_{\tau4} \\ U_{s1} & U_{s2} & U_{s3} & U_{s4} \end{pmatrix} \simeq \begin{pmatrix} c_\odot & s_\odot & \epsilon & \epsilon \\ \epsilon & \epsilon & 1/\sqrt{2} & 1/\sqrt{2} \\ \epsilon & \epsilon & -1/\sqrt{2} & 1/\sqrt{2} \\ -s_\odot & c_\odot & \epsilon & \epsilon \end{pmatrix}, \tag{9}
$$

where $c_\odot \equiv \cos\theta_\odot$, $s_\odot \equiv \sin\theta_\odot$ and $\theta_\odot$ stands for the mixing angle of the SMA MSW solar solution. In this case $c_s \equiv |U_{s1}|^2 + |U_{s2}|^2 \simeq 1$ and the solar neutrino deficit would be accounted for by sterile neutrino oscillations $\nu_e \leftrightarrow \nu_s$ with the SMA MSW solution while the atmospheric neutrino anomaly would be by $\nu_\mu \leftrightarrow \nu_\tau$ oscillations. This scenario is obviously inconsistent with the recent solar neutrino data by the Superkamiokande group, and the argument which has lead to (9) has to be given up.

Fortunately the upper bound on N_ν has become less stringent now and $N_\nu = 4.0$ seems to be allowed. Furthermore, it has been shown recently[33] that the combined analysis of BBN and the recent data by BOOMERanG[34] and MAXIMA-1[35] of the Cosmic Microwave Background prefers higher value of N_ν: $4 \leq N_\nu \leq 13$. Therefore all the four neutrino schemes of type (2+2) and (3+1) seemed to be consistent with the BBN constraint.

On the other hand, it has been pointed out[36] in the two flavor framework that for a certain range of the oscillation parameters neutrino oscillations themselves create asymmetry between ν and $\bar\nu$ and this asymmetry prevents ν_s from oscillating into active neutrinos. Although this analysis has not been generalized to the four neutrino cases, even if the upper bound of N_ν becomes less than 4.0 in the future, it might be still possible to have four neutrino schemes which are consistent with the BBN constraint as well as the solar and atmospheric neutrino data due to possible asymmetry in ν and $\bar\nu$.

6 Conclusions

In this talk I have shown that there are still four neutrino scenarios ((2+2)- as well as (3+1)- schemes) which are consistent with all the experiments and the observations, despite the recent claims by the Superkamiokande group that pure sterile oscillations $\nu_e \leftrightarrow \nu_s$ in the solar neutrinos and pure sterile oscillations $\nu_\mu \leftrightarrow \nu_s$ in the atmospheric neutrinos are disfavored. In particular, the reason that the (2+2)-scheme is consistent with the recent Superkamiokande data is because both solar and atmospheric neutrinos have hybrid oscillations of active and sterile oscillations.

162

Acknowledgments

I would like to thank Yoshitaka Kuno for invitation and the local organizers for hospitality during the workshop. This research was supported in part by a Grant-in-Aid for Scientific Research of the Ministry of Education, Science and Culture, #12047222, #10640280.

References

1. B.T. Cleveland et al., Nucl. Phys. B (Proc. Suppl.) **38**, 47 (1995).
2. Y. Fukuda et al., Phys. Rev. Lett. **77**, 1683 (1996) and references therein.
3. Y. Suzuki, Nucl. Phys. B (Proc. Suppl.) **77**, 35 (1999) and references therein.
4. V.N. Gavrin, Nucl. Phys. B (Proc. Suppl.) **77**, 20 (1999) and references therein.
5. T.A. Kirsten, Nucl. Phys. B (Proc. Suppl.) **77**, 26 (1999) and references therein.
6. Y. Fukuda et al., Phys. Lett. **B335**, 237 (1994) and references therein.
7. R. Becker-Szendy et al., Phys. Rev. **D46**, 3720 (1992) and references therein.
8. T. Kajita and Y. Totsuka, Rev. Mod. Phys. **73**, 85 (2001) and references therein.
9. Y. Suzuki, talk at *19th International Conference on Neutrino Physics and Astrophysics* (Neutrino 2000), Sudbury, Canada, June 16-22, 2000 (http://nu2000.sno.laurentian.ca/ Y.Suzuki/).
10. Y. Fukuda et al., Phys. Rev. Lett. **85**, 3999 (2000).
11. W.W.M. Allison et al., Phys. Lett. **B449**, 137 (1999).
12. C. Athanassopoulos *et al.*, (LSND Collaboration), Phys. Rev. Lett. **77**, 3082 (1996); Phys. Rev. C **54**, 2685 (1996); Phys. Rev. Lett. **81**, 1774 (1998); Phys. Rev. C **58**, 2489 (1998); D.H. White, Nucl. Phys. Proc. Suppl. **77**, 207 (1999).
13. G. Mills, talk at *19th International Conference on Neutrino Physics and Astrophysics* (Neutrino 2000), Sudbury, Canada, June 16–22, 2000 (http://nu2000.sno.laurentian.ca/G.Mills/).
14. L. Borodovsky *et al.*, Phys. Rev. Lett. **68**, 274 (1992).
15. J. Kleinfeller, Nucl. Phys. Proc. Suppl. **85**, 281 (2000).
16. B. Ackar et al., Nucl. Phys. **B434**, 503 (1995).
17. F. Dydak *et al.*, Phys. Lett. B **134**, 281 (1984).
18. N. Okada and O. Yasuda, Int. J. Mod. Phys. **A 12**, 3669 (1997).

19. S.M. Bilenky, C. Giunti and W. Grimus, hep-ph/9609343; Eur. Phys. J. **C1**, 247 (1998).
20. V. Barger, B. Kayser, J. Learned, T. Weiler and K. Whisnant, Phys. Lett. **B489**, 345 (2000).
21. S.M. Bilenky, C. Giunti, W. Grimus and T. Schwetz, Astropart. Phys. **11**, 413 (1999).
22. C. Giunti, M. C. Gonzalez-Garcia and C. Peña-Garay, Phys. Rev. **D62**, 013005 (2000); M. C. Gonzalez-Garcia, talk at 30th International Conference on High-Energy Physics (ICHEP 2000), Osaka, Japan, July 27 – August 2, 2000 (http://ichep2000.hep.sci.osaka-u.ac.jp/scan/0728/pa08/gonzalez_garcia/).
23. O. Yasuda, hep-ph/0006319.
24. G.L. Fogli, E. Lisi and A. Marrone, Phys. Rev. **D63**, 053008 (2001).
25. C. Giunti and M. Laveder, hep-ph/0010009.
26. O.L.G. Peres and A.Yu. Smirnov, hep-ph/0011054.
27. Review of Particle Physics, Particle Data Group, Eur. Phys. J. **C3**, 1 (1998).
28. O. Yasuda, unpublished.
29. O. Yasuda, *New Era in Neutrino Physics* (eds. H. Minakata and O. Yasuda, Universal Academic Press, Tokyo, 1999) p 165 (hep-ph/9809205).
30. D. Dooling, C. Giunti, K. Kang and C.W. Kim, Phys. Rev. D **61**, 073011 (2000).
31. A. Nicolaidis, G. Tsirigoti and J. Hansson, hep-ph/9904415.
32. R. Barbieri and A. Dolgov, Phys. Lett. **B237**, 440 (1990), Nucl. Phys. **B349**, 743 (1991); K. Kainulainen, Phys. Lett. **B244**, 191 (1990); K. Enqvist, K. Kainulainen and M. Thomson, Nucl. Phys. **B373**, 498 (1992), Phys. Lett. **B288**, 145 (1992); X. Shi, D.N. Schramm and B.D. Fields, Phys. Rev. **D48**, 2563 (1993).
33. S. Esposito, G. Mangano, A. Melchiorri, G. Miele and O. Pisanti Phys. Rev. **D63**, 043004 (2001).
34. P. de Bernardis et al., Nature **404**, 955 (2000).
35. A. Balbi et al., Ap. J. **545**, L1 (2000).
36. R. Foot and R.R. Volkas, Phys. Rev. **D55**, 5147 (1997); Astropart. Phys. **7**, 283 (1997); Phys. Rev. **D56**, 6653 (1997).

NEUTRINO OSCILLATIONS AND NEW PHYSICS

SARAH R. NUSS-WARREN, SARAH C. CAMPBELL AND LORETTA M. JOHNSON

Department of Physics, Grinnell College, Grinnell, IA 50112, USA

E-mail:johnsonl@grinnell.edu

We observe that neutrino oscillation experiments are sensitive to flavor change from direct flavor-violating interactions in addition to neutrino oscillations, and we explore the consequences of this "confusion" effect. We show that near detectors at neutrino factories and MINOS could improve present constraints on new physics by more than an order of magnitude. In addition, neutrino factories could observe CP violation from direct interactions.

1   Introduction

Recent neutrino experiments have amassed substantial evidence for solar and atmospheric neutrino flavor change, and the LSND and K2K accelerator experiments have provided tantalizing hints of neutrino flavor change. These accelerator experiments are presently not able to draw strong conclusions, and additional experiments are vital for confirming flavor change and further studying the parameter space. In particular, most models with neutrino mass and mixing also contain direct neutrino interactions not present in the Standard Model, but most studies have not explored these direct interactions [1]. Now that we have strong evidence for neutrino flavor change from solar and atmospheric experiments, we should explore the possibility that direct neutrino interactions provide a sizeable contribution to the flavor change observed in some experiments.

Here we describe our formalism that includes oscillations and direct interactions in an equivalent way [2], and we discuss examples that demonstrate that near-future accelerator experiments may observe flavor change in near detectors from direct interactions. Our formalism provides a model-independent method of studying the synergy of oscillations and direct interactions. Through discussion of the examples below, we will show what kinds of experiments could observe direct interactions and how such experiments could distinguish oscillations alone from the combination of oscillations and direct interactions. We will show representative graphs of parameter space to clarify what future experiments might observe.

2   Formalism

As a clear-cut example, we consider here the cases of neutrinos produced in pion or muon decay with direct flavor violation present at production. These represent two of the most common accelerator neutrino sources, and such flavor violation at the source is possible in a range of models. The resulting equations can be easily applied to a variety of experiments. The general leptonic (L) and semileptonic (S) effective low-energy Lagrangians are then

$$\mathcal{L}^L = 2\sqrt{2}G_F F^{hh'}_{Aijkm}\left(\bar{\ell}_i \Gamma_A P_h U_{ja}\nu_a\right)\left(\bar{\ell}_k \Gamma_A P_{h'} U_{mb}\nu_b\right)^\dagger$$

$$\mathcal{L}^S = 2\sqrt{2}G_F K^h_{Aij}\left(\bar{\ell}_i \Gamma_A P_h U_{ja}\nu_a\right)\left(\bar{d}\Gamma_A\left(\alpha P_L + \beta P_R\right)u\right)^\dagger + h.c.$$

where repeated indices are summed. In these equations the F and K coefficients generalize the Lorentz and flavor structure of the familiar four-Fermi Lagrangians. The Lorentz structure is labeled by $A = S, V, T$ for scalar, vector and tensor, and h = L,R for pseudoscalar and pseudovector, $P_h = 1/2$ ($1\pm\gamma_5$). As usual, the unitary matrix, U_{ja}, relates the mass and flavor states with i, j, k, m labeling the flavor states and a, b the mass states. Note that these flavor states are still the weak eigenstates, but that the F and K coefficients now allow for terms in the Lagrangians representing interactions where lepton flavor is not conserved.

Here we restrict our discussions to extremely relativistic neutrinos and the same $(V - A) \times (V - A)$ Lorentz structure that is present in weak interactions, and so we will omit the label L on our coefficients. We point out that the flavor-change amplitude will depend on the source process – which could include direct interactions – as well as the propagation of the neutrino. In the case of this simple Lorentz structure, we are left with clean expressions for the flavor-change probability with muon ($\mu^+ \to e\nu_e\nu_\mu$ and $\mu_+ \to e\nu_\mu\nu_\mu$) and pion ($\pi^+ \to \mu\nu_\mu$ and $\pi^+ \to \mu\nu_\tau$) decay sources, respectively:

$$P_{e\to\mu} = \left| F^{LL}_{V2k1j}U^*_{ja}e^{-iE_a t}U_{2a}\right|^2$$

$$P_{\mu\to\tau} = \left| K^L_{V2j}U_{ja}e^{-iE_a t}U^*_{ja}\right|^2$$

where repeated indices are summed. We point out that these probabilities were obtained by constructing the amplitude from source, through propagation, to detection, and then squaring and normalizing.

In order to simplify our notation in what follows, we introduce the following definitions:

$$2x \equiv \left(m_b^2 - m_a^2\right)\frac{t}{2E}$$

$$F, K = \tan\psi e^{2i\varphi}$$

where appropriate subscripts will be added when necessary to eliminate misunderstanding. Now for two-flavor mixing with $U_{11} = U_{22} = \cos\theta$ and $U_{12} = -U_{21} = \sin\theta$ we obtain the flavor-change probabilities for muon and pion sources, respectively:

$$P_{e\to\mu} = \left| -2i\sin xe^{-ix}\sin\theta\cos\theta F_{V2211} + \left(1 - 2i\sin xe^{-ix}\sin^2\theta\right)F_{V2212} \right|^2$$

$$+ \left| -2i\sin xe^{-ix}\sin\theta\cos\theta F_{V2111} + \left(1 - 2i\sin xe^{-ix}\sin^2\theta\right)F_{V2112} \right|^2$$

$$P_{\mu\to\tau} = \tan^2\psi_\tau + \sin^2 2\theta\sin^2 x\left(1 - \tan^2\psi_\tau\right) - 2\sin 2\theta\sin x\tan\psi_\tau$$
$$\times\left[\sin(x - 2\varphi_\tau) - 2\cos^2\theta\sin x\cos(2\varphi_\tau)\right]$$

We point out that the muon source leads to a more complicated formula because the undetected neutrinos may be either flavor.

Table 1. Existing upper limits on some direct interaction coefficients.

Coefficient	Process	Constraint
F_{V2111} $(\tan\psi_{ee})$	$\mu \to eee$	8.0×10^{-6}
F_{V2112} $(\tan\psi_{e\mu})$	$\mu^+ e^- \to \mu^- e^+$	6.0×10^{-3}
F_{V2113} $(\tan\psi_{e\tau})$	$\tau \to \mu^+ ee$	9.8×10^{-3}
F_{V2213} $(\tan\psi_{\mu\tau})$	$\tau \to e^-\mu\mu$	1.0×10^{-2}
F_{V2311} $(\tan\psi_{\tau e})$	$\tau \to \mu^- ee$	1.0×10^{-2}
F_{V2312} $(\tan\psi_{\tau\mu})$	$\tau \to e^+\mu\mu$	9.8×10^{-3}
K_{V21} $(\tan\psi_{21})$	$\mu \to e$ conversion	1.5×10^{-6}
K_{V31} $(\tan\psi_{31})$	$\tau \to e\pi^0$	1.5×10^{-2}
K_{V32} $(\tan\psi_\tau)$	NOMAD	1.6×10^{-2}

Constraints on the F and K coefficients (or equivalently on the $\tan\psi$) exist and in some cases are quite strong, as we indicate in Table 1; the phases are unconstrained and in some cases measurable, as noted below. All but one arises from constraints on charged-lepton flavor violation in processes related through SU(2) to neutrino-charged-lepton flavor violation [3]. We note that though the muon decay to three electrons and muon-electron conversion channels are constrained too tightly for near-future experiments to observe direct interactions in the neutrino-related channels, the other constraints are not very restrictive. The recent NOMAD collaboration constraint is from their emulsion search for tau neutrino appearance [4]. It also is of an order of magnitude that near detectors at neutrino factories and MINOS will be able to probe.

3   Neutrino Factories

A detector near a muon decay source, as is planned at Neutrino Factories [5], is ideal for studying some of the least well-constrained parameters. Neutrino Factories provide an abundant source of high energy electron neutrinos (not produced copiously in existing experiments) in addition to muon neutrinos. Sign selection of the muon beam allows only one flavor of neutrino and one flavor of anti-neutrino in the Standard Model. Thus these experiments can look for appearance in two channels at once, switch sign to study the other two channels, and study CP violation by comparing. Near detectors provide particularly strong hope for observing direct interactions since the oscillation probability at a near detector is small.

$P_{e\rightarrow\mu}$ represents the four ways that μ^+ decay can lead to a detected ν_μ: 1) Standard Model interaction (F_{2211}) and ν_e oscillates to ν_μ, 2) new interaction (F_{2212}) produces ν_μ and $\overline{\nu}_\mu$, 3) new interaction (F_{2111}) produces $\overline{\nu}_e$ and ν_e oscillates to ν_μ, and 4) new interaction (F_{2112}) produces $\overline{\nu}_e$ and ν_μ. Generally we can represent this process as $\mu^+ \rightarrow e^+ \nu_\mu \overline{\nu}_x$, where x stands for the unobserved neutrino which may be either flavor. The terms that do not involve oscillation contribute the leading terms, $\tan^2\psi_{\mu\mu}$ and $\tan^2\psi_{e\mu}$, while the other terms involve $\sin^2 2\theta \sin^2 x$. Since F_{2212} is so much less strongly constrained, it may dominate over F_{2112}; in what follows, we do not neglect any terms, even F_{2111}, but the most easily observable behavior is associated with F_{2212}.

The probability for observing the process related through CP, $\mu^- \rightarrow e^- \overline{\nu}_\mu \nu_x$, also contains the relatively unconstrained coefficient, F_{2212}, in the leading term, $\tan^2\psi_{\mu\mu}$. These two probabilities, $P_{e\rightarrow\mu}$ and $P_{\overline{e}\rightarrow\overline{\mu}}$, differ in signs of phases in the term linear in $\tan\psi_{\mu\mu}$, and so they may be out of phase. The remaining two probabilities, $P_{\mu\rightarrow e}$ and $P_{\overline{\mu}\rightarrow\overline{e}}$, involve $\psi_{e\mu}$ in the leading term, $\tan^2\psi_{e\mu}$, which is more strongly constrained. Because of this difference in the leading terms, T violation and CPT violation arising from direct interactions could be large.

In summary, there are tight constraints on all the of parameters associated with $\nu_\mu \rightarrow \nu_e$ and $\overline{\nu}_\mu \rightarrow \overline{\nu}_e$, but $\nu_e \rightarrow \nu_\mu$ and $\overline{\nu}_e \rightarrow \overline{\nu}_\mu$ are relatively unconstrained. For these latter processes, direct interactions could dominate the flavor-change probabilities by an order of magnitude at near detectors; this direct-violation effect would be independent of L/E and so observable at all of the proposed Neutrino Factories. In addition, CP violation from direct interactions could be observable, and T and CPT effects from direct interactions could be large.

4   Main Injector Neutrino Oscillation Search

The MINOS Collaboration [6] will use the more conventional muon neutrino beams from pion decay to study the region of parameter space down to of order 0.01 eV^2 and 0.01 in $sin^2 2\theta$. Sign selection of the pion beam will make the neutrino beam primarily muon neutrinos for π^- and muon anti-neutrinos for π^+. New direct interactions could produce electron neutrinos in the beam, but the probability for this is less than or of order 10^{-6}. We will confine our remarks here to the possibility of tau neutrino appearance in the near detector; in fact, observation of taus would require an emulsion detector that is not presently planned for the MINOS near detector.

Tau neutrinos may be observed at the near detector either because of oscillation or because of direct interactions. The oscillation probability is 2×10^{-8} which will not be measurable at MINOS; this provides the justification for using the near detector to measure the flux of muon neutrinos. As Figure 1 shows, the probability of a direct interaction producing a tau neutrino could be as large as 2×10^{-4} which might be measurable. Although this may not have a large effect on the interpretation of observations at the far detector, observing tau neutrinos at the near detector is an exciting possibility that we urge the MINOS Collaborators to explore.

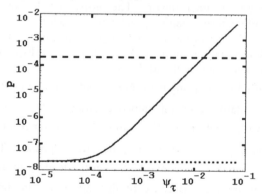

Figure 1. Flavor-change probability as a function of direct interaction strength. The oscillations-only probability (dotted line) is low and constant. The contribution from new direct interactions is potentially big. The dashed line is the NOMAD limit.

Although in principle study of CP violation from direct interactions is possible at the MINOS near detector, in practice the differences in flavor-change probability are small enough to make observation unlikely. Figure 2 shows that for typical parameter choices, the difference in flavor-change probabilities is about 10^{-8}. The reason CP violation is so hard to study is that the same direct interaction coefficient plays the leading role in both equations and only the signs of the phases differ.

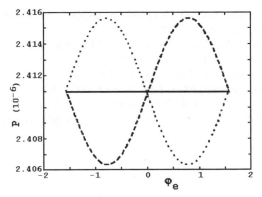

Figure 2. Flavor-change probability as a function of direct interaction phase. The solid line represents oscillations only. The dotted and dashed lines represent flavor-change probability for negative and positive pions, respectively.

Adding an emulsion detector at the MINOS near detector would enable observation of tau neutrinos produced by direct interactions at the neutrino source. If no taus were observed, a new limit on ψ_τ would be established.

5 Conclusions

Our examples demonstrate how experiments designed to study the flavor change of neutrinos from muon and pion beams could observe new direct flavor-violating interactions, particularly at near detectors where oscillation probabilities are extremely small. Neutrino Factories in particular could discover rather large flavor-violating interactions or set new, strong limits on the relevant coefficients. Neutrino Factories also have the opportunity to study T and CPT violation due to direct flavor-violating interactions. In the nearer future, MINOS could observe moderate flavor-violation or set limits perhaps an order of magnitude below their current values if they install an emulsion detector with the near detector at Fermilab. Observation of this flavor-change at the near detector could affect the interpretation of the data collected at Soudan. The recent new limit from NOMAD that improved the limit on ψ_τ by a factor of five should remind us that new flavor-changing interactions could be present at near-future experiments and would affect the interpretation of the flavor-change data.

References

1. V Barger, R. Phillips and K. Whisnant, *Phys. Rev. D* **44** (1991) p. 1629; P. Herzeg and R. Mohapatra, *Phys. Rev. Lett.* **69** (1992) p. 2475; S. Bergmann and

Y. Grossman, *Phys. Rev. D* **59** (1999) p. 093005; M. Gonzalez-Garcia et al., *Phys. Rev. Lett.* **82** (1999) p. 3202.

2. L. Johnson and D. McKay, *Phys. Lett. B* **433** (1998) p. 355; L. Johnson and D. McKay, *Phys. Rev. D* **61** (2000) p. 113007.
3. Particle Data Group, *Eur. Phys. J. C* 15 (2000) p. 1.
4. NOMAD Collaboration, *Phys. Lett. B* 483 (2000) 387.
5. For example, R. Palmer et al, *Proceedings of SnowMass 96*.
6. MINOS Collaboration, http://www.hep.anl.gov/ndk/hypertext/mino_tdr.html (1998).

AMBIGUITIES OF THEORETICAL PARAMETERS AND CP/T VIOLATION IN NEUTRINO FACTORIES

MASAFUMI KOIKE

Institute for Cosmic Ray Research, University of Tokyo, Kashiwa-no-ha 5-1-5,
Kashiwa, Chiba 277-8582, Japan
E-mail: koike@icrr.u-tokyo.ac.jp

TOSHIHIKO OTA

Department of Physics, Kyushu University, Fukuoka 812-8581, Japan
E-mail: toshi@higgs.phys.kyushu-u.ac.jp

JOE SATO

Research Center for Higher Education, Kyushu University,
Ropponmatsu, Chuo-ku, Fukuoka 810-8560, Japan
E-mail: joe@rc.kyushu-u.ac.jp

We study the optimal setup for observation of the CP asymmetry in neutrino factory experiments — the baseline length, the muon energy and the analysis method. First, we point out that the statistical quantity which has been used in previous works doesn't represent the CP asymmetry. Then we propose the more suitable quantity, $\equiv \chi_2^2$, which is sensitive to the CP asymmetry. We investigate the behavior of χ_2^2 with ambiguities of the theoretical parameters. The fake CP asymmetry due to the matter effect increases with the baseline length and hence the error in the estimation of the fake CP asymmetry grows with the baseline length due to the ambiguities of the theoretical parameters. Namely, we lose the sensitivity to the genuine CP-violation effect in longer baseline.

1 Introduction

The observation of the atmospheric neutrino anomaly by Super-Kamiokande [1] provided us with convincing evidence that neutrinos have non-vanishing masses. There is another indication of neutrino masses and mixings by the solar neutrino deficit [2,3,4,5,6].

Assuming three generations of the leptons, we denote the lepton mixing matrix, which relates the flavor eigenstates ($\alpha = e, \mu, \tau$) with the mass eigenstates with mass $m_i (i = 1, 2, 3)$, by

$$U_{\alpha i} = \begin{pmatrix} c_{13}c_{12} & c_{13}s_{12} & s_{13} \\ -c_{23}s_{12} - s_{23}s_{13}c_{12}e^{i\delta} & c_{23}c_{12} - s_{23}s_{13}s_{12}e^{i\delta} & s_{23}c_{13}e^{i\delta} \\ s_{23}s_{12} - c_{23}s_{13}c_{12}e^{i\delta} & -s_{23}c_{12} - c_{23}s_{13}s_{12}e^{i\delta} & c_{23}c_{13}e^{i\delta} \end{pmatrix}_{\alpha i}, \quad (1)$$

where $c_{ij}(s_{ij})$ is the abbreviation of $\cos\theta_{ij}(\sin\theta_{ij})$. Then the atmospheric neutrino anomaly gives an allowed region for $\sin\theta_{23}$ and the larger mass square difference ($\equiv \delta m_{31}^2$). The solar neutrino deficit provides allowed regions for $\sin\theta_{12}(\equiv \delta m_{21}^2)$.

On the other hand, there is only an excluded region for $\sin\theta_{13}$ from reactor experiments [8]. Furthermore there is no constraint on the CP violating phase δ. The idea of neutrino factories with muon storage rings were proposed [9] to determine these mixing parameters, and attracted the interest of many physicists [10,11,12,13,14,15,16,17,18,19,20].

However we have some questions concerning the previous analyses of the CP-violation effect. In many analysis, the muon energy of E_μ is assumed to be rather high. This seems strange since CP/T violation arises as a three generation effect [21,19,22,23]. Indeed we can derive very naively that $E_\mu \sim$ 30GeV, lower by factor 2, is the most efficient for $L = 3000$ km [19] while $E_\mu \sim 50$GeV is often assumed. Furthermore the fake CP-violation effect due to the matter effect[24] increases with baseline length. The ambiguity in the estimate for the fake CP violation increases with baseline length. Taking into account this ambiguity in the analysis, the sensitivity to CP violation will be decreased as baseline length increases. It is unlikely that we can observe the CP-violation effect with such a long baseline. We discuss these problems[23].

2 Statistical quantity

As an experimental setup, we consider that N_μ muons decay at a muon ring. The neutrinos extracted from the ring are detected at a detector if E_ν is larger than a threshold energy E_{th}. The detector has mass M_{detector} and contains N_{target} target atoms. We assume that the neutrino-nucleon cross section σ is proportional to neutrino energy as

$$\sigma = \sigma_0 E_\nu, \tag{2}$$

The expected number of appearance events in the energy bin $E_{j-1} < E_\nu < E_j$ ($j = 1, 2, \ldots, n$) is then given by

$$N_j(\nu_\alpha \rightarrow \nu_\beta; \delta) \equiv \frac{N_\mu N_{\text{target}}\sigma_0}{\pi m_\mu^2} \frac{E_\mu^2}{L^2} \int_{E_{j-1}}^{E_j} E_\nu f_{\nu_\alpha}(E_\nu) P(\nu_\alpha \rightarrow \nu_\beta; \delta) \frac{\mathrm{d}E_\nu}{E_\mu}, \tag{3}$$

where m_μ is the muon mass.

To estimate the sensitivity for the CP-violation effect the following statistics is usually used:

$$\chi_1^2(\delta_0) \equiv \sum_{j=1}^{n} \frac{[N_j(\delta) - N_j(\delta_0)]^2}{N_j(\delta)} + \sum_{j=1}^{n} \frac{[\bar{N}_j(\delta) - \bar{N}_j(\delta_0)]^2}{\bar{N}_j(\delta)} \tag{4}$$

n is the number of bins. Since the CP violation is absent if $\sin \delta = 0$, namely $\delta = 0$ or $\delta = \pi$, we need to check that $N_j(\delta)$ is different from $N_j(\delta_0)$ with $\delta_0 \in \{0, \pi\}$ to insist that CP violation is present.

We can claim that $N_j(\delta)$ is different from $N_j(\delta_0)$ at 90% confidence level, if

$$\chi_1^2 \equiv \min(\chi_1^2(0), \chi_1^2(\pi)) > \chi_{90\%}^2(n) \tag{5}$$

holds. Here $\chi_{90\%}^2(n)$ is the χ^2 value with n degrees of freedom at 90% confidence level.

To see the behavior of χ_1^2, we make use of high energy approximation which is valid for $E_\nu \gtrsim (\delta m_{31}^2 L)/4$:

$$\chi_1^2(\delta_0) \propto E_\mu \frac{J_{/\delta}^2}{A} \left\{ (\cos \delta \mp 1) \left[1 - \frac{1}{3} \left(\frac{a(L)L}{4E_\nu^{\text{peak}}} \right)^2 \right] \right\}^2 \tag{6}$$

Here E_ν^{peak} is the neutrino energy which gives the maximum valu of the initial neutrino flux f_{ν_α} and $a(L)$ is the effective mass square due to matter effect calculated using Preliminary Reference Earth Model(PREM)[27,28]. We find that χ_1^2 is an increasing function of E_μ. Thus we can obtain arbitrary large χ_1^2, and we can seemingly achieve arbitrary high sensitivity to search for the CP-violation effect, by increasing muon energy. Thus the higher energy appears to be preferable to observe the CP-violation effect as long as we employ χ_1^2. It is important, however, to note that χ_1^2 has nothing to do with the imaginary part of the mixing matrix in high energy limit. The CP violation is brought about by the only imaginary part of the mixing matrix, which is proportional to $\sin \delta$ in our parameterization. χ_1^2 is relevant with CP violation through unitarity[22].

Therefore we need to consider a statistical quantity which is sensitive to the imaginary part of the lepton mixings. As such a statistics we consider the following quantity:[14,15]

$$\chi_2^2(\delta_0) \equiv \sum_{j=1}^{n} \frac{[\Delta N_j(\delta) - \Delta N_j(\delta_0)]^2}{N_j(\delta) + \bar{N}_j(\delta)} \tag{7}$$

Here $\Delta N_j(\delta) \equiv N_j(\delta) - \bar{N}_j(\delta)$. It is required

$$\chi_2^2 \equiv \min(\chi_2^2(0), \chi_2^2(\pi)) > \chi_{90\%}^2(n) \tag{8}$$

to claim that CP violation effect is observed.

In the high energy limit

$$\chi_2^2(\delta_0) \propto \frac{L^2}{E_\mu} \frac{J_{/\delta}^2}{A} \left\{ \sin \delta + \frac{1}{3} \frac{a(L)L}{4E_\nu^{\text{peak}}} (2\cos 2\theta_{13} - 1)(\cos \delta \mp 1) \right\}^2 \tag{9}$$

($-$ for $\delta_0 = 0$ and $+$ for $\delta_0 = \pi$), where

$$J_{/\delta} \equiv \frac{\delta m_{21}^2}{\delta m_{31}^2} \sin 2\theta_{12} \sin 2\theta_{23} \sin 2\theta_{13} \cos \theta_{13}, \tag{10}$$

$$A \equiv \sin^2 \theta_{23} \sin^2 2\theta_{13}. \tag{11}$$

To see CP-violation effect is to measure $J_{/\delta} \sin \delta^{25,26}$. In this respect χ_2^2 gives a good standard to observe CP violation.

3 Feasibility of CP violation search in presence of the ambiguities of the parameters

In this section we study the asymmetry with χ_2^2. The values of all theoretical parameters will have ambiguities in practice, and hence we cannot estimate $\Delta N_j(\delta_0)$ precisely. The genuine CP-violation effect will be absorbed into the ambiguity of $\Delta N(\delta_0)$ if the ambiguity of $\Delta N(\delta_0)$ is large. Therefore we must examine whether the CP-violation effect can be absorbed in the ambiguities of the parameters.

Suppose that we use the parameters $\tilde{x}_i \equiv \{\tilde{\theta}_{kl}, \delta \tilde{m}_{kl}^2, \tilde{a}(L)\}$, which are different from the true values $x_i \equiv \{\theta_{kl}, \delta m_{kl}^2, a(L)\}$, to calculate $N_j(\delta_0)$ and $\bar{N}_j(\delta_0)$. We will estimate the fake CP violation due to the matter effect as

$$\Delta \tilde{N}_j(\delta_0) = \tilde{N}_j(\delta_0) - \bar{\tilde{N}}_j(\delta_0), \tag{12}$$

are evaluated from eqs.(3). We then obtain

$$\tilde{\chi}_2^2(\delta_0) \equiv \sum_{j=1}^{n} \frac{[\Delta N_j(\delta) - \Delta \tilde{N}_j(\delta_0)]^2}{N_j(\delta) + \bar{N}_j(\delta)} \tag{13}$$

instead of $\chi_2^2(\delta_0)$. The observed asymmetry $\Delta N_j(\delta)$ consists of the genuine CP-violation effect and the fake one due to the matter effect. We have to subtract the matter effect, but we cannot estimate precisely the fake CP violation $\Delta \tilde{N}_j(\delta_0)$ due to the ambiguities of the parameters. In such a case the sensitivity to CP-violation search gets worse once the ambiguities of the parameters are taken into account, since it is always possible to take $\Delta \tilde{N}_j(\delta_0)$ to satisfy

$$\left| \Delta N_j(\delta) - \Delta \tilde{N}_j(\delta_0) \right| \leq \left| \Delta N_j(\delta) - \Delta N_j(\delta_0) \right|, \tag{14}$$

or equivalently

$$\tilde{\chi}_2^2 \leq \chi_2^2, \tag{15}$$

by adjusting $\tilde{x}_i$'s. We can further argue that we lose more sensitivity as the baseline length gets longer. Let us illustrate the outline described above in detail. The CP asymmetry of probabilities

$$A(\{x_i\},\delta) \equiv P(\nu_\alpha \to \nu_\beta; \{x_i\},\delta) - P(\bar{\nu}_\alpha \to \bar{\nu}_\beta; \{x_i\},\delta) \tag{16}$$

consists of the genuine CP asymmetry $A_{\mathrm{CPV}}(\{x_i\},\delta)$ and the fake one $A_{\mathrm{CPM}}(\{x_i\},\delta)$, so that

$$A(\{x_i\},\delta) = A_{\mathrm{CPV}}(\{x_i\},\delta) + A_{\mathrm{CPM}}(\{x_i\},\delta). \tag{17}$$

We need to subtract $A(\{x_i\},\delta_0)$ from $A(\{x_i\},\delta)$, but instead we subtract $A(\{\tilde{x}_i\},\delta_0)$ due to the ambiguities of the parameters and obtain

$$\tilde{A}_{\mathrm{CPV}}(\delta) \equiv A_{\mathrm{CPV}}(\{x_i\},\delta) + A_{\mathrm{CPM}}(\{x_i\},\delta) - A(\{\tilde{x}_i\},\delta_0). \tag{18}$$

Here A_{CPV} and A_{CPM} can be estimated using high energy approximation as

$$A_{\mathrm{CPM}}(\{x_i\},\delta) \simeq \frac{1}{3}\left[2\sin^2\theta_{23}\sin^2 2\theta_{13}\cos 2\theta_{13}\right.$$
$$\left. + (2\cos 2\theta_{13} - 1)J_{/\delta}\cos\delta\right]\frac{a(L)L}{4E_\nu}\left(\frac{\delta m_{31}^2 L}{4E_\nu}\right)^3 \tag{19}$$

$$A_{\mathrm{CPV}}(\{x_i\},\delta) = \left(\frac{\delta m_{31}^2 L}{4E_\nu}\right)^3 J_{/\delta}\sin\delta. \tag{20}$$

The factor

$$\frac{2}{3}\sin^2\theta_{23}\sin^2 2\theta_{13}\cos 2\theta_{13}\frac{a(L)L}{4E_\nu} \tag{21}$$

in eq.(19) is expected to be much larger than $J_{/\delta}$ in eq.(20) with a long baseline . Thus the ambiguity of the fake CP-violation effect, $A_{\mathrm{CPM}}(\{x_i\},\delta) - A(\{\tilde{x}_i\},\delta_0)$, can absorb the genuine CP-violation effect A_{CPV}, so that $\tilde{A}_{\mathrm{CPV}}$, or equivalently $\tilde{\chi}_2^2$, becomes significantly small. The condition to observe CP-violation effect in 90% confidence level, say, is again given by

$$X_2^2 \equiv \min_{\{\tilde{x}_i\}}\tilde{\chi}_2^2 > \chi_{90\%}^2(n). \tag{22}$$

We present in Figs.1 the required value of $N_\mu M_{\mathrm{detector}}$ obtained from eq.(22) to observe the CP-violation effect in 90% confidence level. All the parameters are assumed to have ambiguities of 10 %. We find that we cannot observe the genuine CP-violation effect when L is larger than 1000 km. We can qualitatively understand it by eqs.(19) and (20). It is seen that

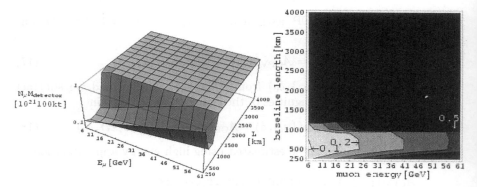

Figure 1. Necessary value of $N_\mu M_{\text{detector}}$ to observe the CP-violation effect as a function of muon energy and baseline length, for $\delta = \pi/2$ and $E_{\text{th}} = 1\text{GeV}$. Here $(\sin\theta_{13}, \sin\theta_{23}, \sin\theta_{12}, \delta m_{31}^2, \delta m_{21}^2) = (0.1, 1/\sqrt{2}, 0.5, 3\times10^{-3}\text{eV}^2, 10^{-4}\text{eV}^2)$ and $a(L)$ is calculated using PREM. The ambiguities of the theoretical parameters are assumed to be 10 %. Hence these graphs are obtained using not χ_2^2 but X_2^2. The sensitivity to the genuine CP asymmetry is lost in long baseline region such as $L \gtrsim 1250\text{km}$ as we estimate in eq.(25).

$$\frac{A_{\text{CPV}}}{A_{\text{CPM}}} = 3\frac{J_{/\delta}\sin\delta}{2\sin^2\theta_{23}\sin^2 2\theta_{13}\cos 2\theta_{13} + (2\cos 2\theta_{13} - 1)J_{/\delta}\cos\delta}\frac{4E}{a(L)L} \quad (23)$$

is a decreasing function of L, which means that the sensitivity to the CP violation is lost as the baseline length gets larger. The condition on L is roughly estimated by $A_{\text{CPV}}/A_{\text{CPM}} \gtrsim 1$, or

$$L \lesssim \frac{4E}{a(L)}\frac{3J_{/\delta}\sin\delta}{2\sin^2\theta_{23}\sin^2 2\theta_{13}\cos 2\theta_{13} + (2\cos 2\theta_{13} - 1)J_{/\delta}\cos\delta}. \quad (24)$$

For the parameters used in Fig.1,

$$L \lesssim 1250\text{km}. \quad (25)$$

4 Summary and Discussion

We discussed the optimum experimental setup and the optimum analysis to see the CP violation effect.

We examined how to analyze the data of experiments to confirm the naive estimation. We studied with two statistical quantities, χ_1^2 (eq.(5)) and χ_2^2 (eq.(8)). Usually χ_1^2 is used in analyses of neutrino factories. We can test by this whether the data can be explained by the hypothetical data calculated assuming no CP-violation effect. We saw, however, that this quantity is sensitive has information for mainly the CP conserved part of the oscillation probability in high energy region. Hence we concluded that it is difficult to measure the CP violation by using this quantity. On the other hand, we can test with χ_2^2 whether the asymmetry of oscillation probabilities of neutrinos and antineutrinos exists. We have seen that χ_2^2 is sensitive to the CP violating part of the oscillation probability, and thus it is suitable quantity to measure the CP violation.

Then we investigated the influence of the ambiguities of the theoretical parameters on χ_2^2. Since the matter effect causes the difference of the oscillation probabilities between neutrinos and antineutrinos, we have to estimate the fake asymmetry to search for the CP violation effect. However, we will always "overestimate" the fake CP violation due to the ambiguities of the theoretical parameters, and hence we will always estimate the genuine CP-violation effect too small. The matter effect increases as baseline length increases, and we will lose the sensitivity to the asymmetry due to the genuine CP-violation effect in longer baseline such as several thousand km.

References

1. Super-Kamiokande Collaboration, Y. Fukuda *et al.*, Phys. Rev. Lett. **81** (1998) 1562; Phys. Lett. **B433** (1998) 9; Phys. Lett. **B436** (1998) 33; Phys. Rev. Lett. **82** (1999) 2644.
2. GALLEX Collaboration, W. Hampel *et al.*, Phys. Lett. B **447** (1999) 127.
3. SAGE Collaboration, J. N. Abdurashitov *et al.*, astro-ph/9907113.
4. Kamiokande Collaboration, Y. Suzuki, Nucl. Phys. B (Proc. Suppl.) **38** (1995) 54.
5. Homestake Collaboration, B. T. Cleveland *et al.*, Astrophys. J. **496** (1998) 505.
6. Super-Kamiokande Collaboration, Y. Fukuda *et al.*, Phys. Rev. Lett. **82** (1999) 1810; *ibid.* **82** (1999) 2430.

7. G. L. Fogli, E. Lisi, D. Marrone and G. Scioscia, Phys. Rev. D **59** (1999) 033001; G. L. Fogli, E. Lisi, D. Montanino and A. Palazzo Phys. Rev. D **62** (2000) 013002.

8. M. Apollonio *et al.*, Phys. Lett. **B420**, 397 (1998); Phys. Lett. B466, 415 (1999).

9. S. Geer, Phys. Rev. **D57**, 6989 (1998), erratum *ibid.* **D59** (1999) 039903.

10. V. Barger, S. Geer and K. Whisnant, Phys. Rev. **D61** (2000) 053004.

11. A. Cervera , A. Donini, M.B. Gavela, J. J. Gomez Cadenas, P. Hernandez, O. Mena and S. Rigolin, Nucl. Phys. **B579** (2000) 17.

12. V. Barger, S. Geer, R. Raja and K. Whisnant, Phys. Rev. **D62** (2000) 013004; Phys. Lett. **B485** (2000) 379; hep-ph/0007181.

13. V. Barger, S. Geer, R. Raja and K. Whisnant, Phys. Rev. **D62** (2000) 073002;

14. A. De Rujula, M. B. Gavela and P. Hemandez, Nucl. Phys. **B547**, 21 (1999); A. Donini, M. B. Gavela, P. Hemandez and S. Rigolin, Nucl. Phys. **B574** (2000) 23.

15. K. Dick, M. Freund, M. Lindner, and A. Romanino, Nucl. Phys. **B562** (1999) 29; A. Romanino, Nucl. Phys. **B574** (2000) 675; M. Freund, P. Huber and M. Lindner, Nucl. Phys. **B585** (2000) 105.

16. M. Freund, M. Lindner, S.T. Petcov and A. Romanino, Nucl. Phys. **B578** (2000) 27.

17. Neutrino Factory and Muon Collider Collaboration (D. Ayres *et al.*), physics /9911009; C. Albright *et al.* hep-ex/0008064.

18. M. Campanelli, A. Bueno and A. Rubbia, hep-ph/9905240; A. Bueno, M. Campanelli and A. Rubbia, Nucl. Phys. **B573** (2000) 27; Nucl. Phys. **B589** (2000) 577.

19. M. Koike and J.Sato, Phys. Rev. **D61** (2000) 073012; J. Sato, hep-ph/0006127.

20. O. Yasuda, hep-ph/0005134.

21. M. Kobayashi and T. Maskawa, Prog. Theor. Phys. 49, 652 (1973).

22. J. Sato, hep-ph/0008056.

23. M. Koike, T. Ota and J. Sato, hep-ph/0011387.

24. L. Wolfenstein, Phys. Rev. **D17** (1978) 2369; S. P. Mikheev and A. Yu. Smirnov, Sov. J. Nucl. Phys. **42** (1985) 913.

25. J. Arafune and J. Sato, Phys. Rev. D **55**, 1653 (1997).

26. J. Arafune, M. Koike and J. Sato, Phys. Rev. D **56**, 3093 (1997); erratum *ibid.* **60**, 119905 (1999).

27. A. M. Dziewonski and D. L. Anderson, *Phys. Earth Planet. Inter.* **25**, 297 (1981)

28. T. Ota and J. Sato, hep-ph/0011234.

NEUTRINO FACTORY PLANS AT CERN

J. A. RICHE

For the Neutrino Factory Working Group

CERN, 1211,Geneve 23,
SWITZERLAND
E-mail: alain.riche@cern.ch

The considerable interest raised by the discovery of neutrino oscillations and recent progress in studies of muon colliders has triggered interest in considering a neutrino factory at CERN. This paper explains the reference scenario, indicates the other possible choices and mentions the R&D that are foreseen.

1 Introduction

CERN is considering a neutrino factory [1,2,3] as a possible option for the post LHC era. In such a facility, electron-neutrinos and muon-neutrinos are created in equal amounts by high-energy muon decay. The detectors are located close to the source and at long (~700 km) or very long (~3000 km) distances. The problem is the production of intense muon beams and their rapid acceleration to the final energy. However, a very solid conceptual background exists, since muon colliders were studied in the 90's [4,5,6]. For less demanding neutrino production, the muon beam emittance can be higher than for a collider, and energy lower than 50 GeV is sufficient. The front-end of a muon collider, with a more modest ionization cooling section, thus inspires the design of most scenarios for neutrino factories [7], except the FFAG scenario. In the case of CERN, the availability of LEP cavities, the considerable experience with 40 and 80 MHz cavities, as well as site considerations have favoured a scenario based on low energy (2.2 GeV) protons. The factory consists of a proton driver, a target area, a muon front-end, a sequence of muon accelerators, and a muon decay ring.

2 The Proton Driver

A low-energy, high-power system consisting of a 2.2 GeV linac, a fixed-field accumulator and a compressor produces the 10^{16} protons/s required for having 10^{14} muons/s in the muon decay ring. The time structure and bunch length of the proton beam is appropriate for acceleration and storage of the resulting muon beam.

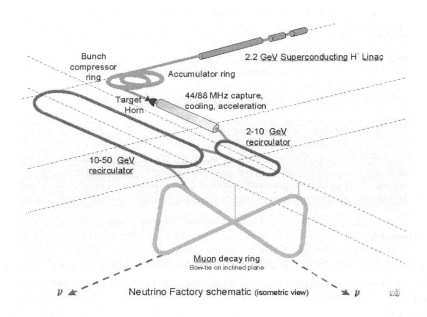

Figure 1. A possible layout for a Neutrino Factory.

Since the CERN PS injectors (50 MeV Linac and 1.4 GeV Booster) cannot deliver the beam power required for a Neutrino Factory, a Superconducting H⁻ Linac (SPL) has been proposed [8,9] with an average beam power of 4 MW, an energy of 2.2 GeV, and a repetition rate of 75 Hz. It uses 116 of the 352 MHz superconducting LEP cavities for the acceleration from 1 GeV to 2.2 GeV, and similar cavities with lower β for acceleration to 1 GeV. The layout is shown in Fig.2. Regarding the acceptable radiation level, the large aperture of the LEP RF cavities permits the transmission of the halo to the end of the linac, where scraping and dumping of the halo will be done before injection into the rings.

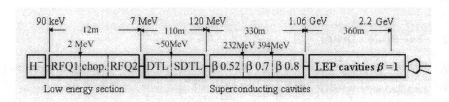

Figure 2 Layout of the Superconducting Linac.

The SPL accelerates H⁻ (rather than protons, to allow charge exchange injection using a foil stripping into the following circular machine) in bursts of 2.2 ms duration. The muon lifetime at 50 GeV is about 1 ms, and each burst should be separated by at least 2 ms. The repetition frequency of 75 Hz of the SPL fulfills this condition. The beam should be accumulated to obtain further:

- a burst duration compatible with the circumference of the muon storage ring (i.e.<2 km). The burst length will be reduced by a factor of 660 by accumulation in a ring of suitable circumference, but the bunch length will be increased.
- a bunch duration compatible with the RF accelerating the muons. This reduction of bunch duration is done at the expense of an increased energy spread of the proton beam, which is unimportant for muon production.

The functions of accumulation and bunching are separated into two rings (Fig.3) of 3.3 μs revolution time, corresponding to the length of the old Intersection Storage Rings (ISR) tunnel at CERN [10].

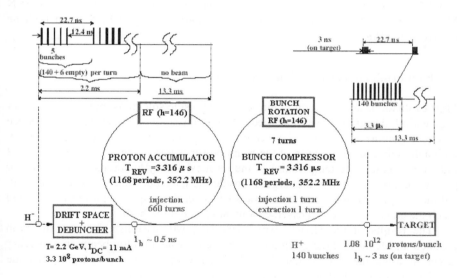

Figure 3 Layout of the Proton Accumulator and Compressor.

The 44 MHz RF captures the 352 MHz SPL bunches, in groups of 8, and there are 146 of them around the ring. For 2.2 ms SPL burst length, the charge exchange injection is made in 660 turns. At the end of the injection, the central part of the 44 MHz buckets is filled in a quasi-uniform way by the filamentation.

At the end of the accumulation, the 146 bunches are fast ejected and single-turn injected into the bunch compressor. This ring has the same circumference and also 44 MHz RF. The bunch compression is performed in 7 turns by phase rotation; the resulting bunch length on target is 3 ns.

Other scenarios have been studied for different proton energies, in collaboration with Rutherford Appleton Laboratory [11]: at 5 GeV/50 Hz and at 15 GeV/25 Hz, both with 2 boosters and two rings; and a 180 MeV/56 mA H⁻ linac, similar to the European Spalliation Source design. A 30 GeV/8 Hz machine [12] injecting into the SPS ring above transition energy γ_t would provide naturally short bunches because of the high γ_t. It had been also studied for the case higher energy and lower repetition rate would be preferred.

In addition to its use for the Neutrino Factory, the SPL could provide a higher brilliance beam for the PS and improved beam characteristics for other experiments such as: CERN Neutrinos to Gran Sasso (CNGS) operated by SPS, Anti-proton Decelerator (AD), Neutrons Time of Flight (n-TOF), and Isotope Separator On Line (ISOLDE). The 800 m SPL could be installed along the boundary of the CERN Meyrin site, close to the accumulator and compressor rings which could both be housed in the ISR tunnel.

3   Pion Production, Collection and Decay

Most evaluations are based on liquid jet targets inside solenoids, but no satisfactory engineering design has yet emerged. Active research is underway to produce practical systems of targets and collectors. The target could be a liquid mercury jet. Although only a small fraction of the beam power is lost in the 30 cm long target (2 radiation lengths), the temperature of the jet may rise by hundreds of degrees. For pion collection efficiency, the target is inside the magnetic field or very close to it. All the charged particles emitted forward whatever their sign, spiral in the field and reach the end of it if their transverse momentum is not too high.

The results of the HARP experiment [13] will establish data for the pion production. Meanwhile the generation of particles from a proton beam on a mercury target in a magnetic field can be calculated with codes for hadronic cascade simulations, such as FLUKA, with the data presently available.

An example of these calculations is shown in Fig.4, for a 2.2 GeV beam on a 30 cm long, 0.75 cm radius Hg target. The target is inside the 20-T magnetic field of a 15-cm diameter solenoid The calculated total production per incident proton in forward direction is ~0.1 π^+ (average energy 310 MeV) and 0.075 π^- (average energy 288 MeV). This simulation [14] predicts that the forward production of π^+ increases by a factor 15 if the energy is increased by a factor 8 from 2 to 16 GeV.

With 10^{23} proton per year on the target, the total neutron dose in the solenoid is ~10^{11} Gy per year from neutrons and is ~$1.4 \ 10^{10}$ Gy per year from γ.

The collection of π⁺ in an interval of ± 100 MeV around a central kinetic energy of ~200 MeV, corresponds to 40% of the forward yield, the momentum acceptance being imposed by the accelerating devices used downstream.

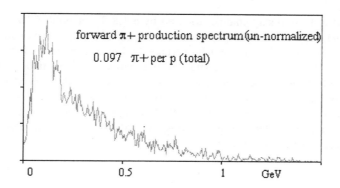

Figure 4 Example of π+ production (not normalized), by a 2.2 GeV proton on a 30 cm mercury target

Other options are also being considered for the collection of pions:

- a pulsed magnetic horn [7]. Particles of one sign are bent towards the axis; those of the opposite sign are deflected away (see Fig.5)
- a cylindrical liquid mercury target carrying a high pulsed current along its length, with a radial distribution of circulating mercury at both ends for cooling and shock-wave damping. This arrangement is not yet designed [15].
- a rotating solid target followed by focusing devices [7].

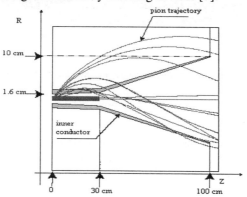

Figure 5 Principle of focusing by the transverse magnetic fields of a horn.

4   MUON FRONT END

This is the section from the exit of the muon collector up to the entrance to the recirculator, according to the reference scenario [16,17].

4.1   Phase Rotation

At the end of a 30 m decay channel focused by a 1.8 T solenoid, 99% of the 300 MeV/c (or 191 MeV kinetic energy) π^+ have decayed. The bunching is partially lost, as the high-energy particles of a bunch overtake the low energy ones of the previous bunch (Fig.6). But with 22.7 ns separation, this happens for momenta lower than ~200 MeV/c (100 MeV kinetic energy) which are not captured. The reduction by a factor 2 of the energy spread is obtained by phase rotation in the 44 MHz RF cavities, giving a kinetic energy of 200 ± 50 MeV. The RF phase is set to decelerate the particles with the higher energy arriving first, and accelerate the particles with the lower energy, which arrive later.

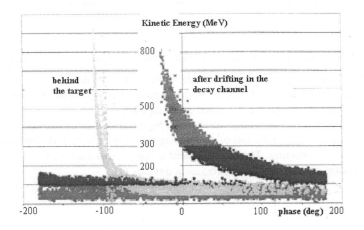

Figure 6 Energy versus phase of a bunch after the target and after the 30 m. decay channel.

4.2   Ionization Cooling

Cooling reduces the normalized emittance in order to fit the acceptance of the recirculators. It results from absorption of the energy in tanks filled with H_2, so longitudinal and transverse momenta are both reduced. In each cooling unit, RF re-acceleration restores the longitudinal momentum, not the transverse one and the angle of the particle with the axis diminishes. The multiple Coulomb scattering counteracts the reduction of the angle. But within the present parameters, this effect is weak. The optics of the cooling line is setup to maintain a high divergence and thus make the cooling efficient. Fig.7 shows the reduction of the momentum by the energy loss (a), the increase of the transverse momentum by scattering (b), the re-acceleration (c), the rotation of the ellipse of emittance keeping the divergence high (d).

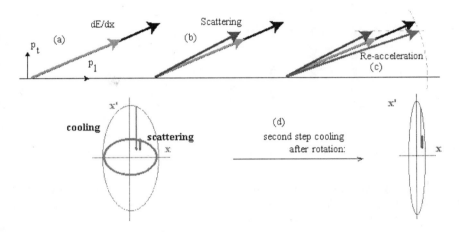

Figure 7 Ionization cooling, RF re-acceleration and rotation of the ellipse of emittance.

4.3   Linear Acceleration

Linacs accelerate the beam to 2 GeV in the following steps, including cooling:

- Cooling by degrading the energy and associated restoring of the longitudinal momentum is provided by 4 RF cells, each 1 m long, 2 MV/m (44 MHz), 30 cm bore radius, with a solenoid housed in each cell, followed by one degrader unit of 0.25 m long. A reduction of the emittance by a factor 1.4 is expected.
- Acceleration and bunching to reach 280 MeV kinetic energy.
- Cooling and associated re-acceleration are repeated again at 88 MHz with cavities of 15 cm bore radius, a field of 4 MV/m and built-in solenoids. A reduction of the emittance by a factor 4 is expected.
- Acceleration at 88 MHz (4 MV/m), then 176 MHz (10 MV/m), up to 2 GeV.

The muon losses expected from production through cooling and acceleration to 2 GeV are shown in Fig. 8:

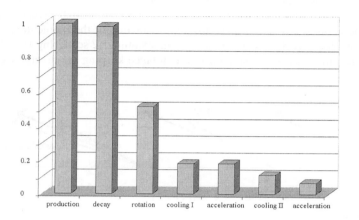

Figure 8 Muon budget in the RF scheme, up to 2 GeV

A cooling experiment is to be studied at CERN using a PS proton beam at 26 GeV and a target area. An energy absorber associated with energy-compensating RF cavities could reduce the emittance by 5 to 10% [21].

5 ACCELERATION FROM 2 GeV TO 50 GeV

Rapid Cycling Synchrotrons, recirculators and Fixed Field Accelerating Gradient accelerators have been proposed for accelerating the muons to full energy. The CERN scenario uses recirculators, but an alternative solution, with FFAG rings, could avoid bunch rotation and cooling. FFAG accelerators have a very large acceptance, and the magnetic field, being constant with time, allows for fast acceleration. In principle, the radial extension of vacuum chamber and magnets is limited by a magnetic field varying with radius R as:

$$B = B_o (R / R_o)^n$$

with large B_o and n values.

Successive sector magnets with fields of alternating sign increasing with radius give a net curvature toward the machine centre. In a genuine FFAG as proposed at KEK [18,19,20], the closed orbits for different energies are similar and the number of betatron oscillations is the same for all energies. For machines with only a few turns, crossing resonance lines is not disastrous and keeping Q constant not a strict condition.

In CERN scenario, two recirculating accelerators RLA1 and RLA2 increase the muon beam energy to 10, then to 50 GeV [22]. Each consists of two linacs, with 352 MHz superconducting cavities of 7 MV/m field, and arcs for re-circulating the beam 4 times. The design is similar with that of the ELFE study [23] at CERN. By choosing the gradients of the quadrupoles in proportion to the beam energy, a constant phase advance per cell is provided for the transverse focusing in the linac in the first pass. The arcs are achromatic. At the ends of the linacs, spreaders separate the beam in four vertically stacked arcs. At the end of the arcs, combiners take these beams back to the linacs. The lattice of the arc with spreaders and combiners is isochronous. The beam acceleration occurs at the crest of the RF wave, and the muon energy spread, initially equal to 12.5%, is adiabatically damped.

Table 1. Recirculating linear accelerators.

	RLA1	RLA2	
Energy	2 to 10	10 to 50	GeV
Number of turns	4	4	
Length of linacs	2×340	2×1900	m
Circumference	806	4442	m

6   MUON STORAGE RING

About 77% of the muons survive acceleration in the RLAs. They are then injected into the muon storage ring at 50 GeV, with two high-beta long straight sections aligned with the neutrino detectors at distances of 700 and 3000 km [24]. Neutrinos produced in the arcs are lost for physics, so arc length is minimized.

A short straight section in the arc could be used for short base line detector. Both triangular and bow tie shape decay rings have been investigated (see Table 2). The ring is equipped with RF to preserve the bunching of the muon beams, necessary for detection with position monitors and avoiding the depolarization of the muons. The normalized rms emittance is 1.7 π mm rad; the aperture limit is at 3 σ, the relative rms momentum spread is 0.5%. The muon beam divergence contributes quadratically to the divergence of the neutrinos, which is

$\sigma'_\nu \approx 1/\gamma_\mu \approx 2\ 10^{-3}$.

Choosing the contribution of the divergence of the muon beam to be ten times less, $\sigma'_\mu \approx (1/10)\ \sigma'_\nu \approx 0.2$ mrad, the normalized muon beam divergence is $\gamma_\mu\,\sigma'_\mu \approx 0.1$. The aperture limitation of the long straight section of 3σ limits the muon divergence to 0.6 mrad.

The average power density due to the electrons from the decays is 150 W/m. Due to the contribution of the long straight sections much higher power density is dissipated at the entrances of the arcs.

Table 2. Muon Storage Ring (bow-tie shape)

Energy	50	GeV
Length of straight section	576	m
Vertical extension of ring	150	m
Circumference	2008	m
Bending angle of arcs	2×250	deg.

The layout of the neutrino factory according to the preliminary study at CERN is shown Fig.9. The machine tunnels will be constructed in the molasse layer in the vicinity of CERN.

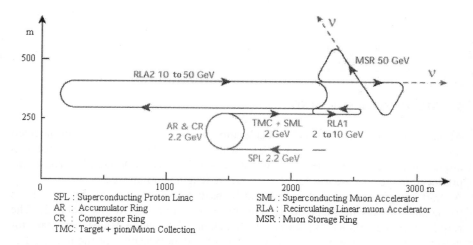

Figure 9 Layout out of Neutrino Factory [24]

7 CONCLUSIONS

The CERN scenario of the Neutrino Factory is conditioned by the investigation of limited regions of the parameter space. A Superconducting linac, an accumulator and a compressor deliver a 2.2 GeV, 5 MW primary beam with the convenient temporal structure. Pions are collected close to the maximum of the production avoiding the need of a too large momentum acceptance. Compression and relatively modest energy cooling give the emittance and time structure necessary for the acceleration to 50 GeV and the decay of more than 10^{21} muon per year in the straight sections of a bow-tie decay ring. Other options are being studied in the US and Japan. The final choice of the parameters should emerge from the comparison of the various projects.

8 ACKNOWLEDGEMENTS

The present work is the result of the effort of the Neutrino Factory Working Group at CERN. The help of other laboratories, in particular RAL, GSI, CEA; and of members of the American Neutrino Factory and Muon Collider Collaboration is gratefully acknowledged. Helpful contacts with KEK are very much appreciated.

References:

1. B. Autin, A. Blondel, J. Ellis (Eds.), Prospective Study of Muon Storage Rings at CERN, CERN 99-02, ECFA 99-197 (April 2000).
2. H. Haseroth, Status of Studies for a Neutrino Factory at CERN, European Particle Accelerator Conference, June 2000, Vienna.
3. B. Autin, J.P. Delahaye, R. Garoby, H. Haseroth, K. Hubner, C.D. Johnson, E. Keil, A. Lombardi, H.L. Ravn, H, Schonauer, A. Blondel, The CERN Neutrino Factory Working Group, Status Reports and Working Plans, CERN-NUFACT-Note 28, (August 2000).
4. R. Palmer, C.D. Johnson, E. Keil, A Cost-Effective Design for a Neutrino Factory, CERN SL-99-070, (November 1999).
5. N. Holtkamp, D. Finley (Eds.), A Feasibility Study of a Neutrino Source Based on a Muon Storage Ring, FERMILAB pub-00/08-E, (June 2000).
6. C.M. Ankenbrandt et al., Status of Muon Collider Research and Development and Future Plans, Phys. Rev. ST Accel. Beams 2, 08001 (August 1992).
7. Neutrino Factories based on Muon Storage Rings, ('νFACT 99'), Proceedings of the ICFA/ECFA Workshop, LYON 5-9 July 1999, NIM, A, Vol. 451, No 1
8. M. Vretenar (Editor), Report on the Study Group on a Superconducting Proton Linac as a PS Injector, CERN/PS 98-063 (RF-HP), (December 1998).
9. R. Garoby, M. Vretenar, Status of the Proposal for a Superconducting Proton Linac at CERN, CERN/PS/99-064 (RF), (November 1999).
10. B. Autin, R. Cappi M. Chanel, J. Gareyte, R. Garoby, M. Giovanozzi, H. Haseroth, M. Martini, E. Metral, D. Mohl, K. Schindl, H. Schonauer, I. Hofmann, C. Prior, G. Rees, S. Koscielniak, Design of a 2 GeV Accumulator-

Compressor for a Neutrino Factory, European Particle Accelerator Conference, June 2000, Vienna, and NUFACT-Note 31

11. C.R. Prior, Neutrino Factories Studies at RAL, European Particle Accelerator Conference, June 2000, Vienna.

12. H. Schonauer, B. Autin, R. Cappi, M. Chanel, J. Gareyte, R. Garoby, M. Giovannozzi, H. Haseroth, M. Martini, E. Metral, W. Pirkl, K. Schindl, C. Prior, G. Rees, I. Hofmann, Y. Senichev, A Slow-Cycling Proton Driver for a Neutrino Factory, European Particle Accelerator Conference, June 2000, Vienna.

13. Spokesperson: F. Dydak, Proposal to Study Hadron Production for the Neutrino Factory and for the Atmospheric Neutrino Flux, Harp PS-214, CERN-SPSC/99-35, SPSC/P315 (November 1999).

14. J.P. Amand, CERN, oral communication, November 2000.

15. B. Autin, CERN, oral communication, December 2000.

16. A. Lombardi, A 40-80 MHz System for Phase Rotation and Cooling, CERN NUFACT-Note 20, (August 2000).

17. G. Franchetti, S. Gilardoni, P. Gruber, K. Hanke, H. Haseroth, E.B. Holzer, D. Kuchler, A. Lombardi, R Scrivens, Phase Rotation, Cooling and Acceleration of Muons Beams: a Comparison of Different Approach, LINAC 2000 Conference, August 2000, Monterey

18. M. Aiba, K. Koba, S. Machida, Y. Mori, R. Muramatsu, C. Ohmori, I.Sakai, Y. Sato, A. Takagi, R. Ueno, T. Yokoi, M. Yoshimoto, Development of a FFAG Proton Synchrotron, European Particle Accelerator Conference, June 2000, Vienna.

19. Y. Mori, The Japanese Neutrino Factory, this workshop.

20. Y. Kuno, PRISM machine, this workshop.

21. NFWG special meeting dedicated to muon cooling and selected diagnostics instrumentation, CERN, 13 Dec. 2000.

22. E. Keil. Neutrino Factories Accelerator Facilities, CERN SL/2000-043, CERN-NUFACT/Note 33, and Neutrino 2000, Sudbury, Canada, June 2000.

23. E. Keil, ELFE at CERN, A 25 GeV C.W. Electron accelerator, CERN SL-2000-022-AP, (July 2000).

24. E. Keil, A 50 GeV Muon Storage Ring Design, CERN/SL/2000-013 AP, CERN NUFACT-Note 26 (May 2000).

NEUTRINO FACTORY IN JAPAN

YOSHIHARU MORI

KEK, High Energy Accelerator Research Organization,
Oho 1-1, Tsukuba-shi, Ibaraki-ken, Japan,
E-mail: yoshiharu.mori@kek.jp

The Japanese scheme of a neutrino factory based on fixed-field alternating(FFAG) gradient accelerators is presented in this paper. The FFAG based neutrino factory has several advantages compared with the linear accelerator based designs in U.S. and Europe.

1 Introduction

One of the most striking next-generation facilities of particle physics is a neutrino factory having high energy muon storage ring with long straight muon decay sections in it. This type of facility is capable of studying 3×3 neutrino MNS mixing matrix in lepton sector intensively. The number of muon decays in the muon storage ring is aimed to be more than 1×10^{20} muon decays/one straight section/year. The energy range of muons is $20-50$ GeV, The accelerator complex of the neutrino factory considered in Japan consists of several sequential rings of the fixed-field alternating gradient (FFAG) accelerators[1], which is significantly different from the others in U.S. and Europe. The idea of FFAG originated from a Japanese physicist, Ohkawa, in 1953. Since then, except several electron models of FFAG built for MURA project in U.S. in the mid 60s, it has not been made by the most recent year when the KEK accelerator group constructed the small POP (proof of principle) proton model in 2000. This machine was successfully operated. The conventional neutrino factory scheme, so called "PJK" scenario, which is based on the linear accelerators and muon storage ring, has been proposed. A high accelerating gradient and small total length of the accelerator minimizes beam loss caused by muon decay, but requires that the rf frequency used in the linear accelerator system becomes relatively high. The typical rf frequency range utilized in this scheme is several 100 MHz. Moreover, a small total length of the linear accelerator system also helps to reduce the cost of the accelerator. The muon survival for various accelerating field gradients when the muons are accelerated from 300 MeV/c to 20 GeV/c is shown in Fig.1.

In the linear accelerator based scenario, the rf frequency used in the accelerating cavity has to be several 100 MHz at least to achieve at least more than 5MV/m, otherwise, the total cost becomes substantial. The disadvantage of

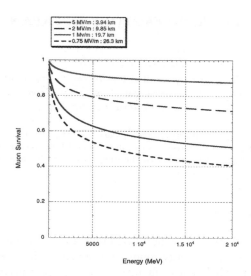

Figure 1. Muon survival during acceleration from 300 MeV/c to 20 GeV/c for various accelerating gradients and fractional distances along the machine.

the high frequency system is its small beam. Thus, in this case, muon beam cooling becomes essential. Ionization cooling consists of a number of energy degrading media between the rf accelerating cavities, and seems to be a possible solution. To make cooling efficient, the accelerating field gradient of the rf cavity has to be large and also a high frequency rf system whose frequency range is more than 100MHz is unavoidable. Since the initial pions, and the product muons have a large energy spread, phase rotation before cooling is also required to decrease the energy spread. Even with this, the muon beam intensity after cooling could drop substantially. If a ring accelerator can be adopted to muon acceleration, this limitation becomes modest. Many turns for acceleration in the same ring using the same accelerating system helps to reduce the total size of the accelerator and the total construction cost. As can be seen in Fig.1, even when the accelerating gradient is only 1MV/ m, the muon survival during acceleration up to 20GeV/c is still more than 50 %, which should not be so painful. The ordinary synchrotron is obviously inadequate for accelerating muons. The magnetic field in an ordinary synchrotron must increase during acceleration and the ramping rate cannot be fast enough to compete with the muon lifetime. The maximum magnetic field ramping rate for a conventional steel electro-magnet is limited by eddy current loss to less than about 200 T/sec. Thus, we consider that a static magnetic field must be used in ring accelerators for muon acceleration.

The FFAG (fixed-field alternating gradient) accelerator seems to be adequate for accelerating muons to high energy. The FFAG is a strong focusing type of synchrotron having a static magnetic field. The concept of the FFAG accelerator was proposed by Ohkawa in 1953. In the early 1960s, this type of accelerator was widely studied and small electron models were developed mostly in North America under the MURA project. However, no practical FFAG had ever been built until recently.

In 1999, development of the proton model of the FFAG accelerator (POP model) was started at KEK and the first proton beam acceleration was successfully achieved in June of 2000. Contrary to electron acceleration, acceleration of a heavy particle such as the proton in an FFAG accelerator is rather difficult, because the rf accelerating system must have a frequency modulation that matches the varying beam revolution time, and have a large accelerating field gradient. A new type of broadband rf cavity using a soft magnetic alloy (MA cavity) has been developed at KEK. The bandwidth of this type of rf cavity is very broad because of its small Q-value (Q<1). The attainable rf field strength becomes very large compared with that of ferrite which has been widely used as the inductive material for the proton synchrotron A big advantage of the FFAG accelerator for accelerating short lived particles such as muons is that the beam guiding magnetic field is static. The acceleration time can be short enough to eliminate the particle decay if the rf voltage is large enough.

Another advantage of the FFAG accelerator is that it has a large acceptance for both transverse and longitudinal directions. In the FFAG accelerators, there are two different types from the beam dynamics point of view; one is the scaling type and the other the non-scaling type. In the scaling type of FFAG accelerator, the beam orbit scales for different energies, which means that the betatron tunes for both horizontal and vertical directions are always constant during acceleration. This is the so-called "zero-chromaticity" condition.

2 Proton driver

The 50 GeV proton synchrotron of the joint project between KEK and JAERI, which will begin construction April , 2001 as a 6-year term project, is considered to be the proton driver for the future neutrino factory. The planned 50-GeV proton synchrotron consist of a 400-MeV proton linear accelerator (400-MeV linac) as an injector, a 3-GeV rapid cycling synchrotron as a booster

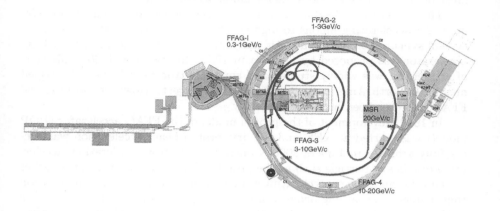

Figure 2. 50-GeV proton synchrotron complex at Tokai site including a FFAG based Neutrino Factory.

and a 50-GeV proton synchrotron (main ring). The accelerators will be constructed at the south site of JAERI-Tokai.

Fig.2 shows schematic layout of the accelerator complex of the joint project at the JAERI Tokai site. The main ring is to accelerate protons from 3 GeV to 50 GeV. The expected beam intensity in the main ring is 3.3×10^{14} ppp and the repetition rate is about 0.3 Hz. The 50-GeV protons are extracted by slow and fast extraction schemes into two experimental areas: one is for experiments using secondary beams (K, antiproton, etc.) and primary beams by slow extraction, and the other is for the neutrino oscillation experiments by fast extraction. When operated in a slow extraction mode, the average current and duty factor, which is defined as the fraction of a cycle when a beam is available, are 15 μA and 0.20, respectively.

Four batches from the booster are injected into the main ring when the main ring stays at a low field. Then, 8 buckets out of 10 are filled with beams, and the main ring starts acceleration while three other facilities start to use 3-GeV beams directly from the booster.

Protons are accelerated from 3 GeV to 50 GeV in the main ring. At the top energy of the 50- GeV main ring, γ is 54.3. In a conventional way of designing a lattice using a regular FODO cell, the transition γ approximately equals the horizontal betatron tune (n_x). In a machine of this scale, because

Table 1. Main Parameters of the accelerator complex

Parameter	present	upgrade
Proton Energy	50 GeV	50 GeV
Protons/pulse	3.3×10^{14}	8.2×10^{14}
Pulse Rate	0.37 Hz	0.66 Hz
Beam Power	1.0 MW	4.4 MW
$\mu-$acceptance (μ/p)	0.21	0.21
$\mu-$survival (N_μ/N_{source})	0.52	0.52

n_x is about 20-30, it is difficult to avoid the transition energy in the regular FODO lattice. Although transition energy crossing techniques have been developed in many operational proton synchrotrons, it is favorable to place the transition energy, where the phase focusing becomes zero, well above the maximum energy, avoiding the instabilities and associated beam losses. Thus, an imaginary transition γ lattice, in which the momentum compaction factor is negative, is employed.

The average beam current of 15.6 μA for slow extraction and 19.6 μA for fast extraction at the first stage will be increased in future. There are several upgrading options. Roughly speaking, two major paths should be taken: one is increase of repetition rate and the other is increase of particles per pulse. Although the repetition rate of the main ring at the beginning is 0.29 Hz for slow extraction mode and 0.37 Hz for fast extraction mode, the lattice magnets themselves are designed so that a higher repetition rate such as 0.51 Hz operation for slow extraction and 0.79 Hz for fast extraction will be possible. That pushes up the average current to roughly 26.8 μA and 41.6 μA, respectively. In this case, the electric power required for exciting the lattice magnets increases and becomes almost doubled as shown below. Although the main ring is not a space charge limited synchrotron at the design particles per pulse in terms of space charge tune shift, special care is necessary if we need to increase the number of particles. One of the options to increase the number of particles is to use barrier buckets at injection. Capturing with barrier buckets decreases the local line density at injection so that the tune shift becomes less. Another advantage of using barrier buckets for injection is that we can inject as many booster batches as we want in contrast with bunch to bucket transfer. The number of batch is rather limited by transverse space charge effects or longitudinal momentum spread. If a higher repetition rate and barrier bucket injection with 10 beam batches are adopted simultaneously in future operation, the average current increases up to 59.4 μA for slow extraction mode and 86.9 μA for fast extraction mode. In case of the neutrino

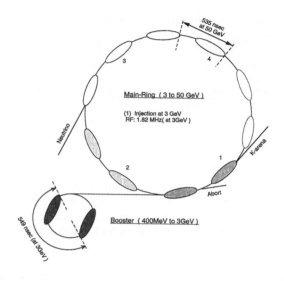

Figure 3. Typical machine cycle structure

factory, the duty factor for slow extraction mode, which means a portion of the flat-top time duration in one main-ring cycle, becomes roughly 33 %. More descriptions on power supply upgrading and barrier buckets injection are below.

2.1 Bunch structure for the FFAG based neutrino factory

In the linac based neutrino factory, in order to reduce the beam emittance efficiently by ionization cooling, phase rotation in longitudinal phase space to decrease the momentum spread of the muon beams becomes essential. Thus, the bunch length of the beam from the proton driver should be rather small in the linac based neutrino factory. The required bunch length in this case is about 3ns or less in rms size. On the other hand, in the FFAG based scenario, the requirement of the bunch length is much more modest compared with this because the longitudinal acceptance of the FFAG using a low frequency rf system is relatively large. The expected bunch width from the proton beam should be even 6 ns or more in rms size for the FFAG based neutrino factory because of its large longitudinal acceptance of the FFAG ring as described later. In ordinary operation, the rf frequency at 50 GeV is about 2 MHz and the bunching factor at 50 GeV becomes about 0.038. Thus, the rms bunch length at 50 GeV is approximately 6 ns in ordinary operation, which is almost the same as required in the FFAG based neutrino factory. This means that no special treatment to the bunch shortening is necessary for the 50GeV proton driver in our FFAG based neutrino factory. If we need in future a bunch shortening for some reasons such as muon polarization, we may take the several schemes for this purpose. The possible schemes to make a shorter bunch are described below.

(1) **Double harmonic number**: The peak beam current at 50 GeV reaches almost 200 A, This is not an easy value to compensate for its beam loading effects on the rf system. It is preferable not to exceed the peak beam current from 200 A, which means that the bunching factor should be around 0.038 even when we shorten the bunch width to half of the ordinary one. To shorten a bunch width of less than 6 ns in rms size while keeping the same bunching factor, we increase the harmonic number twice at the top energy of the 50 GeV ring. It is rather hard to change all of the harmonic numbers from the 3-GeV booster to 50-GeV main ring. The new harmonic number becomes 20. In order to realize this, a second harmonic rf system at the top energy of the 50 GeV ring should be introduced, and bunch manipulation with de-bunching and re-bunching could be applied. As described before, broadband rf cavities with soft magnetic alloy (MA cavity) are used for acceleration in the 50 GeV ring. The Q-value of this type of cavity can be rather low and controllable by cutting cores. In case of the 50 GeV ring, any Q values between 1 and 10 can be set by varying the core spacing. The second harmonic rf cavities using the same material are installed in the 50 GeV ring for increasing the bunching factor at beam injection to reduce the transverse space charge effect. These rf cavities may also be used for the bunch shortening. The main items to be examined from the beam dynamics view in this scheme are microwave instability during the debunching and coupled bunch instability after rebunching. As for the microwave instability, the stability condition would be the same as that of the ordinary operation, which is $Z/n < 2W$, because the bunching factor is the same. Low Q cavities are used in the rf acceleration of the 50 GeV ring. Provided the low Q cavities are also utilized for doubling the harmonic number, the coupled-bunch instability can be avoided.

(2) **Other schemes**: There are a couple of methods to achieve this : (1) rf amplitude jump, (2) rf phase jump, and (3) γ_t manipulation. Method (1) is a common one, however, very high gradient and low frequency rf cavities

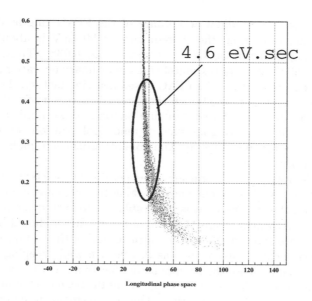

Figure 4. Longitudinal phase space 4.6 eV.sec

are necessary and transient beam loading effect has to be cured carefully. Methods (2) and (3) has been preliminarily tried at the HIMAC synchrotron and the KEK 12-GeV PS, respectively. It seems to work at the intensity level of 10^{12} ppp for the method (3), however, further intensive studies including simulations and experiments should be done.

3 FFAG based neutrino factory

In Fig. 2, a conceptual schematic layout of the FFAG based neutrino factory with the 50 GeV proton driver at JAERI Tokai site is also presented. Since the practical momentum range from injection to extraction in the FFAG accelerator is about 3-4 times, there are four FFAG rings for acceleration of muons from the momentum of 0.3 GeV/c to 20 GeV/c in this scheme. The basic beam parameters for each one are summarized in Table 2.

The particle distribution of the initial pions, and the product muons generated by a short bunched 50 GeV proton beam in the longitudinal phase space. The particle distribution of the initial pions and the product muons in the longitudinal phase space after the captured solenoid when the 50 GeV pro-

Table 2. Main Parameters of FFAG accelerator complex.

momentum(GeV/c)	0.3 to 1 (normal)	0.3 to 1 (super)	1 to 3 (normal)	1 to 3 (super)	3 to 10	10 to 20
average radius(m)	21	10	80	30	90	200
number of sector	32	16	64	32	64	120
k value	50	15	190	63	220	280
beam size at extraction(mm)	170×55	143×55	146×41	115×25	93×17	194×34

ton driver described above is used are shown in Fig.3. In this case, the bunch length of the proton beam from the 50 GeV proton driver is assumed to be 6 nsec in rms value. As can be seen in the figure, the particles having central momentum and momentum spread of 300 MeV/c and ±50%, respectively, are well within the area of 4.6 eV·sec.

This size of longitudinal acceptance can be realized by a low frequency rf accelerating system having an accelerating field gradient of 1 MV/m. One of the advantages in using a low frequency rf system is its large longitudinal acceptance. The typical longitudinal acceptance with such a low frequency rf system would be several eV·sec or more. Such a low accelerating field gradient can be realized with a rather low frequency rf accelerating system. For example, in the anti-proton decelerator (AD) at CERN, the 9 MHz rf cavity has achieved a field gradient of about 0.8 MV/m in burst mode operation. Thus, a ring accelerator is practically the only scheme possible for muon acceleration with a low frequency rf system.

The horizontal acceptance of the FFAG accelerator is very large because of this feature and normally exceeds 10 π mm·rad in real phase space. The momentum acceptance is also very large and a beam having a large momentum spread of more than ± 50% can be accelerated. Thus, both muon cooling and, accordingly, phase rotation should not be necessary. This may become a kind of "brute" force option for muon acceleration in the neutrino factory.

In order to accelerate muon beams from 0.3 GeV/c to 20 GeV/c, four FFAG rings, which are connected in cascade, have been designed. The first ring accelerates muons from 0.3 GeV/c to 1 GeV/c, followed by the second one of 1 GeV/c to 3 GeV/c, the third one of 3 GeV/c to 10 GeV/c, and the final one of 10 GeV/c to 20 GeV/c. The momentum ratio of injection and extraction for each ring is less than 3, which is modest to give a small beam excursion of about 0.5m.

A scaled radial sector type of FFAG with a triplet focusing, which is same for the POP machine, is applied to each ring. There are several advantages

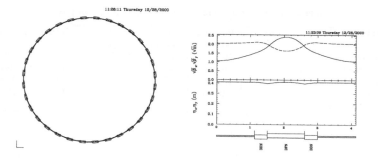

Figure 5. Typical beam parameters of FFAG ring which accelerates muons from 0.3 GeV/c to 1.0 GeV/c. Raidus = 10 m, a number of sectors = 16, magnetic field = 2.8 T.

in triplet configuration compared to the other radial and spiral types. One is field cramping which is expected between adjacent focusing and defocusing magnets. Second, the length of each straight section becomes large because one focusing and two half defocusing magnets are combined together to make one multi function magnet. Finally , the lattice functions has mirror symmetry at the center of a straight section. In our FFAG based neutrino factory, muons are accelerated thorough four FFAG rings from the momentum of 0.3 GeV/c to 20 GeV/c as described above. In order to examine the longitudinal particle motions in the beam acceleration through these four FFAG rings using superconducting magnets, particle tracking simulation has been carried out for. The assumed initial longitudinal emittance and the maximum momentum spread at injection are 4.6 eV·sec and ± 50%, respectively. In this simulation, the averaged rf accelerating field strength of about 1 MV/m is assumed for all of the rings. The simulation results are summarized in Fig 5.As can be seen from these results, the particles are accelerated up to the final energy without having serious problems and the momentum spread is reduced to less than ± 5%.

4   Storage Ring

A storage ring is designed and main parameters are listed in Table 3. It has two of approximately 300 m straight sections. At the straight section, beam size is enlarged and the rms divergence of beams becomes 0.92. That satisfies the condition of

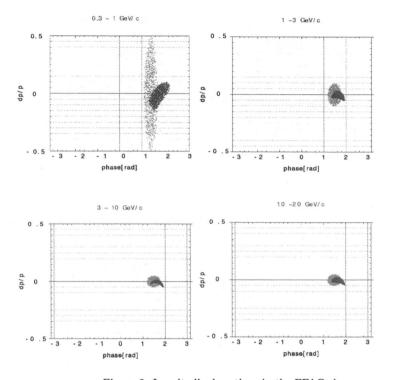

Figure 6. Longitudinal motions in the FFAG rings

$$D_{beam} < 1/(5\lambda) \qquad (1)$$

where λ is a relativistic Lorentz factor.

5 Summary

In this FFAG based neutrino factory, the expected muon intensity after acceleration exceeds more than 6×10^{20} muons/year with the 50 GeV and 0.75 MW proton driver and about 2×10^{20} muon decays/year in the muon storage ring can be realized. If the 50 GeV proton driver is upgraded to reach the beam power of 4.4 MW as described below, the more than 1×10^{21} muon

Table 3. Muon storage ring design parameters and constraints

Parameters	unit	values
Storage ring geometry		racetrack
Storage ring energy	GeV	20
$\varepsilon(100\%)$(normalized)	mm·mrad	$30,000\pi$
$\Delta p/p$	%	1
maximum poletip field	T	<5.0
arc cell phase advance	dep	90

decays/year becomes possible.

References

1. NufactJ Working Group: "A Feasibility Study of a Neutrino Factory in Japan", Feb. 2001.

TESTING NEUTRINO PROPERTIES AT LONG BASELINE EXPERIMENTS AND NEUTRINO FACTORIES

SANDIP PAKVASA

Department of Physics & Astronomy, University of Hawaii, Honolulu, HI 96822, USA

E-mail: pakvasa@phys.hawaii.edu

It is shown that explanations of atmospheric neutrino anomaly other than $\nu_\mu - \nu_\tau$ oscillations (e.g. decay, decoherence and $\nu_\mu - \nu_\tau - \nu_{KK}$ mixing) can be tested at future facilities. Stringent tests of CPT invariance in neutrino oscillations can also be performed.

1 Introduction

In this talk I would like to (a) review some non-oscillatory explanations for the atmospheric neutrino data, and how they can be distinguished from the conventional oscillatory explanations in future neutrino experiments; and (b) describe briefly the strong limits that can be placed on CPT violation both from existing data as well as future experiments especially at neutrino factories.

I should mention on a personal note that having been involved in proposing neutrino oscillations for explaining the atmospheric neutrino anomaly from the beginning[1] it remains my favorite scenario and the one "most likely to succeed." But the point of this exercise is to establish oscillations as the unique solution and to rule the others out.

Let us first consider the following three explanations offered for atmospheric neutrino observations: (i) flavor changing neutral currents (FCNC) of massless neutrinos[2]; (ii) Flavor violation by gravity or violation of Lorentz Invariance[3]; (iii) neutrino decay with mixing with the δm^2 in oscillations is large so that the survival probability is given by[4]

$$P_{\mu\mu} = sin^4\theta + cos^4\theta \exp(-\alpha L/E) \tag{1}$$

In all the above three cases, good fits were obtained for contained events and even partially contained (multi-GeV) events, but as soon as the higher energy (especially the up-coming muon) events are included the fits are much poorer[5].

In all these three cases, the reason that the inclusion of high energy up-coming muon events makes the fits poorer is very simple. The upcoming muons come from much higher energy $\nu'_\mu s$ and although there is some suppression, it is less than what is observed for lower energy events at the same L (zenith angle). This is in accordance with expectations from conventional oscillations. The energy dependence in the above three scenarios is different and fails to account for the data. In the FCNC case there is no energy dependence and so the high energy $\nu'_\mu s$ should have been equally depleted, in the FV Gravity (or Lorentz invariance violation) at high energies the oscillations should average out to give uniform 50% suppression and in the decay A scenario due to time dilation the decay is suppressed and there is hardly any depletion of $\nu'_\mu s$.

2   Neutrino Decay, Decoherence and Extra Dimensions

Decay

If neutrinos do have masses and mixings; then in general, in addition to oscillating, the heavier neutrinos will decay to lighter ones via flavor changing processes. The only questions are (a) whether the lifetimes are short enough to be interesting and (b) what are the dominant decay modes. To be specific, let us assume that neutrino masses are at most of order eV[6].

For eV neutrinos, the only radiative mode possible is $\nu_i \rightarrow \nu_j + \gamma$. From the existing bounds on neutrino magnetic moments, indirect (but model independent) bounds on the decay rates for this mode can be derived: $10^{-11}s^{-1}, 10^{-17}s^{-1}$ and $10^{-19}s^{-1}$ for ν_τ, ν_μ and ν_e respectively[6].

The only possibility for fast invisible decays of neutrinos seems to lie with majoron models. Models of this kind which can give rise to fast neutrino decays and satisfy all bounds have been discussed by Valle, Joshipura and others[7]. These models are unconstrained by μ and τ decays which do not arise due to the $\Delta L = 2$ nature of the coupling. The I=I coupling is constrained by the bound on the invisible Z width; and requires that the Majoron be a mixture of I=1 and I=0. The couplings of ν_μ and ν_e (g_μ and g_e) are constrained by the limits on multi-body π, K decays $\pi \rightarrow \mu\nu\nu\nu$ and $K \rightarrow \mu\nu\nu\nu$ and on $\mu - e$ university violation in π and K decays[8].

Granting that models with fast, invisible decays of neutrinos can be constructed, can such decay modes be responsible for any observed neutrino anomaly?

We assume a component of ν_α, i.e., ν_2, to be the only unstable state, with a rest-frame lifetime τ_0, and we assume two flavor mixing, for simplicity:

$$\nu_\alpha = cos\theta\nu_2 + sin\theta\nu_1 \qquad (2)$$

with $m_2 > m_1$.

If δm^2 is so small that the cosine term is 1, then the survival probability
is

$$P_{\mu\mu} = (sin^2\theta + cos^2\theta \exp(-\alpha L/2E))^2 \qquad (3)$$

Turning to this decay scenario, consider the following possibility[9]. The
three states ν_μ, ν_τ, ν_s (where ν_s is a sterile neutrino) are related to the mass
eigenstates ν_2, ν_3, ν_4 by the approximate mixing matrix

$$\begin{pmatrix} \nu_\mu \\ \nu_\tau \\ \nu_s \end{pmatrix} = \begin{pmatrix} \cos\theta & \sin\theta & 0 \\ -\sin\theta & \cos\theta & 0 \\ 0 & 0 & 1 \end{pmatrix} \begin{pmatrix} \nu_2 \\ \nu_3 \\ \nu_4 \end{pmatrix} \qquad (4)$$

and the decay is $\nu_2 \to \bar{\nu}_4 + J$. The electron neutrino, which we identify with
ν_1, cannot mix very much with the other three because of the more stringent
bounds on its couplings[8], and thus our preferred solution for solar neutrinos
would be small angle matter oscillations.

The decay model of Equation (3) above gives a very good fit to the Super-
K data[10] with a minimum $\chi^2 = 33.7$ (32 d.o.f.) for the choice of parameters

$$\tau_\nu/m_\nu = 63 \text{ km/GeV}, \quad \cos^2\theta = 0.30 \qquad (5)$$

and normalization $\beta = 1.17$.

The best fits of the two models (viz. $\nu_\mu - \nu_\tau$ oscillations and decay) are
of comparable quality. The reason for the similarity of the results obtained
in the two models can be understood by looking at Fig. 1, where I show the
survival probability $P(\nu_\mu \to \nu_\mu)$ of muon neutrinos as a function of L/E_ν for
the two models using the best fit parameters. In the case of the neutrino decay
model (thick curve) the probability $P(\nu_\mu \to \nu_\mu)$ monotonically decreases from
unity to an asymptotic value $sin^4\theta \simeq 0.49$. In the case of oscillations the
probability has a sinusoidal behaviour in L/E_ν. The two functional forms
seem very different; however, taking into account the resolution in L/E_ν, the
two forms are hardly distinguishable. In fact, in the large L/E_ν region, the
oscillations are averaged out and the survival probability there can be well
approximated with 0.5 (for maximal mixing). In the region of small L/E_ν
both probabilities approach unity. In the region L/E_ν around 400 km/GeV,
where the probability for the neutrino oscillation model has the first minimum,
the two curves are most easily distinguishable, at least in principle.

For the atmospheric neutrinos in Super-K, two kinds of tests have been
proposed to distinguish between ν_μ-ν_τ oscillations and ν_μ-ν_s oscillations and
carried out. One is based on the fact that matter effects are present for
ν_μ-ν_s oscillations[11] but are nearly absent for ν_μ-ν_τ oscillations[12] leading to
differences in the zenith angle distributions due to matter effects on upgoing

neutrinos[13]. In our case since the mixing is $\nu_\mu - \nu_\tau$ no matter effect is expected; and hence the recent Super-K results[14] are in accord with expectations of this decay model. The other test is based on the neutral current rate (as measured via production or multi-ring events) which is unaffected in $\nu_\mu - \nu_\tau$ oscillations but reduced in $\nu_\mu - \nu_s$ oscillations[15]. In the decay model, the neutral current rate is affected but less than for $\nu_\mu - \nu_s$ mixing and the current Super-K results which disfavour $\nu_\mu - \nu_s$ mixing cannot yet rule out the decay possibility.

Long-Baseline Experiments

The survival probability of ν_μ as a function of L/E is given in Eq. (1). The conversion probability into ν_τ is given by

$$P(\nu_\mu \to \nu_\tau) = \sin^2\theta\cos^2\theta(1 - e^{-\alpha L/2E})^2 . \tag{6}$$

This result differs from $1 - P(\nu_\mu \to \nu_\mu)$ and hence is different from $\nu_\mu - \nu_\tau$ oscillations. Furthermore, $P(\nu_\mu \to \nu_\mu) + P(\nu_\mu \to \nu_\tau)$ is not 1 but is given by

$$P(\nu_\mu \to \nu_\mu) + P(\nu_\mu \to \nu_\tau) = 1 - \cos^2\theta(1 - e^{-\alpha L/E}) \tag{7}$$

and determines the amount by which the predicted neutral-current rates are affected compared to the no oscillations (or the $\nu_\mu - \nu_\tau$ oscillations) case. Fig. 2 shows the results for $P(\nu_\mu \to \nu_\mu)$, $P(\nu_\mu \to \nu_\tau)$ and $P(\nu_\mu \to \nu_\mu) + P(\nu_\mu \to \nu_\tau)$ for the decay model and compare them to the $\nu_\mu - \nu_\tau$ oscillations, for both the K2K[16] and MINOS[17] (or the corresponding European project[18]) long-baseline experiments, with the oscillation and decay parameters as determined in the fits above.

The K2K experiment, already underway, has a low energy beam $E_\nu \approx$ 1–2 GeV and a baseline $L = 250$ km. The MINOS experiment will have 3 different beams, with average energies $E_\nu = 3$, 6 and 12 GeV and a baseline $L = 732$ km. The approximate L/E_ν ranges are thus 125–250 km/GeV for K2K and 50–250 km/GeV for MINOS. The comparisons in Figure 2 show that the energy dependence of ν_μ survival probability and the neutral current rate can both distinguish between the decay and the oscillation models. ICANOE and especially MONOLITH can also test for the oscillation dip[19].

Decoherence[20]

There are several different possibilities that can give rise to decoherence of the neutrino beam. An obvious one is violation of quantum mechanics[21], others are unknown (flavor specific) new interactions with environment etc. Quantum gravity effects are also expected to lead to effective decoherence[22,23].

The density matrix describing the neutrinos no longer satisfies the usual equation of motion:

$$\dot{\rho} = -i[H, \rho] \tag{8}$$

but rather is modified to

$$\dot{\rho} = -i[H, \rho] + D(\rho) \tag{9}$$

Imposing reasonable conditions[24] on $D(\rho)$ it was shown by Lisi et al.[20] that the ν_μ survival probability $P_{\mu\mu}$ has the form:

$$P_{\mu\mu} = cos^2 2\theta + sin^2 2\theta \, e^{-\gamma L} cos\left(\frac{\delta m^2 L}{2E}\right). \tag{10}$$

where γ is the decoherence parameter. If δm^2 is very small ($\delta m^2 L/2E \ll 1$), this reduces to

$$P_{\mu\mu} = cos^2 2\theta + sin^2 2\theta \, e^{-\gamma L} \tag{11}$$

If $\gamma = \alpha/E$ with α constant, then an excellent fit to the Super-K data can be obtained with $\theta = \pi/4$ and $\alpha \sim 7.10^{-3}$ GeV/Km. (If gamma is a constant, no fit is possible and gamma can be bounded by 10^{-22} GeV). The fits to Super-K data are shown in ref. 20. and they are as good as the decay or $\nu_\mu - \nu_\tau$ oscillations[25]. The shape of $P_{\mu\mu}$ as a function of L/E is very similar to the decay case as shown in Fig. 1.

Large Extra Dimensions

Recently the possibility that SM singlets propagate in extra dimensions with relatively large radii has received some attention[26]. In addition to the graviton, right handed neutrino is an obvious candidate to propagate in some extra dimensions. The smallness of neutrino mass (for a Dirac neutrino) can be linked to this property of the right handed singlet neutrino[27]. The implications for neutrino masses and oscillations in various scenarios have been discussed extensively[28,29]. I focus on one particularly interesting possibility for atmospheric neutrinos raised by Barbieri et al[30]. The survival probability $P_{\mu\mu}$ is given by

$$P_{\mu\mu}(L) = | \Sigma_{i=1}^3 V_{\alpha i} V_{\alpha i}^* A_i(L) |^2 . \tag{12}$$

where

$$A_i(L) = \Sigma_{n=0}^\infty U_{on}^{(i)^2} \, exp(i\lambda_n^{(i)^2} L/2ER^2) \tag{13}$$

where n runs over the tower of Kaluza-Klein states, $\lambda^{(i)}{}_n/R^2$ are the eigenvalues of the mass-squared matrix and $U_{on}^{(i)} (\approx 1/\pi^2 \xi^2)$ are the matrix elements of the diagonalizing unitary matrix.

An excellent fit to the atmospheric neutrino data can be obtained with the following choice of parameters:

$$\xi_3 = m_3 R \sim 3, 1/R \sim 10^{-3} eV, V_{\mu 3}^2 \approx 0.4. \tag{14}$$

The fit to Super-K data is shown in Ref. 30 and obviously it is as good as oscillations. This case corresponds to ν_μ oscillating into ν_τ and a large number(about 25) of Kaluza-Klein states. Because of the mixing with a large number of closely spaced states, the dip in oscillations gets washed out and $P_{\mu\mu}$ looks very much like the decay model as shown in the Fig. 1.

3   CPT Violation in Neutrino Oscillations[32]

Consequences of CP, T and CPT violation for neutrino oscillations have been written down before[32]. We summarize them briefly for the $\nu_\alpha \to \nu_\beta$ flavor oscillation probabilities $P_{\alpha\beta}$ at a distance L from the source. If

$$P_{\alpha\beta}(L) \neq P_{\bar{\alpha}\bar{\beta}}(L), \qquad \beta \neq \alpha, \tag{15}$$

then CP is not conserved. If

$$P_{\alpha\beta}(L) \neq P_{\beta\alpha}(L), \qquad \beta \neq \alpha, \tag{16}$$

then T-invariance is violated. If

$$P_{\alpha\beta}(L) \neq P_{\bar{\beta}\bar{\alpha}}(L), \qquad \beta \neq \alpha, \tag{17}$$

or

$$P_{\alpha\alpha}(L) \neq P_{\bar{\alpha}\bar{\alpha}}(L), \tag{18}$$

then CPT is violated. When neutrinos propagate in matter, matter effects give rise to apparent CP *and* CPT violation even if the mass matrix is CP conserving.

The CPT violating terms can be Lorentz-invariance violating (LV) or Lorentz invariant. The Lorentz-invariance violating, CPT violating case has been discussed by Colladay and Kostelecky[33] and by Coleman and Glashow[34].

The effective LV CPT violating interaction for neutrinos is of the form

$$\bar{\nu}_L^\alpha b_\mu^{\alpha\beta} \gamma_\mu \nu_L^\beta, \tag{19}$$

where α and β are flavor indices. We assume rotational invariance in the "preferred" frame, in which the cosmic microwave background radiation is isotropic (following Coleman and Glashow[34]).

$$m^2/2p + b_0, \tag{20}$$

where b_0 is a hermitian matrix, hereafter labeled b.

In the two-flavor case the neutrino phases may be chosen such that b is real, in which case the interaction in Eq. (19) is CPT odd. The survival probabilities for flavors α and $\bar{\alpha}$ produced at $t = 0$ are given by[34]

$$P_{\alpha\alpha}(L) = 1 - \sin^2 2\Theta \sin^2(\Delta L/4)\,, \tag{21}$$

and

$$P_{\bar{\alpha}\bar{\alpha}}(L) = 1 - \sin^2 2\bar{\Theta} \sin^2(\bar{\Delta} L/4)\,, \tag{22}$$

where

$$\Delta \sin 2\Theta = \left|(\delta m^2/E)\sin 2\theta_m + 2\delta b e^{i\eta} \sin 2\theta_b\right|\,, \tag{23}$$

$$\Delta \cos 2\Theta = (\delta m^2/E)\cos 2\theta_m + 2\delta b \cos 2\theta_b\,. \tag{24}$$

$\bar{\Delta}$ and $\bar{\Theta}$ are defined by similar equations with $\delta b \to -\delta b$. Here θ_m and θ_b define the rotation angles that diagonalize m^2 and b, respectively, $\delta m^2 = m_2^2 - m_1^2$ and $\delta b = b_2 - b_1$, where m_i^2 and b_i are the respective eigenvalues. We use the convention that $\cos 2\theta_m$ and $\cos 2\theta_b$ are positive and that δm^2 and δb can have either sign. The phase η in Eq. (23) is the difference of the phases in the unitary matrices that diagonalize δm^2 and δb; only one of these two phases can be absorbed by a redefinition of the neutrino states.

Observable CPT-violation in the two-flavor case is a consequence of the interference of the δm^2 terms (which are CPT-even) and the LV terms in Eq. (19) (which are CPT-odd); if $\delta m^2 = 0$ or $\delta b = 0$, then there is no observable CPT-violating effect in neutrino oscillations. If $\delta m^2/E \gg 2\delta b$ then $\Theta \simeq \theta_m$ and $\Delta \simeq \delta m^2/E$, whereas if $\delta m^2/E \ll 2\delta b$ then $\Theta \simeq \theta_b$ and $\Delta \simeq 2\delta b$. Hence the effective mixing angle and oscillation wavelength can vary dramatically with E for appropriate values of δb.

We note that a CPT-odd resonance for neutrinos ($\sin^2 2\Theta = 1$) occurs whenever $\cos 2\Theta = 0$ or

$$(\delta m^2/E)\cos 2\theta_m + 2\delta b \cos 2\theta_b = 0\,; \tag{25}$$

similar to the resonance due to matter effects [35,36]. The condition for antineutrinos is the same except δb is replaced by $-\delta b$. The resonance occurs for neutrinos if δm^2 and δb have the opposite sign, and for antineutrinos if they have the same sign. A resonance can occur even when θ_m and θ_b are both small, and for all values of η; if $\theta_m = \theta_b$, a resonance can occur only if $\eta \neq 0$. If one of ν_α or ν_β is ν_e, then matter effects have to be included.

If $\eta = 0$, then

$$\Theta = \theta\,, \tag{26}$$

$$\Delta = (\delta m^2/E) + 2\delta b\,. \tag{27}$$

In this case a resonance is not possible. The oscillation probabilities become

$$P_{\alpha\alpha}(L) = 1 - \sin^2 2\theta \sin^2 \left\{\left(\frac{\delta m^2}{4E} + \frac{\delta b}{2}\right) L\right\}\,, \tag{28}$$

$$P_{\bar{\alpha}\bar{\alpha}}(L) = 1 - \sin^2 2\theta \sin^2 \left\{ \left(\frac{\delta m^2}{4E} - \frac{\delta b}{2} \right) L \right\} . \tag{29}$$

For fixed E, the δb terms act as a phase shift in the oscillation argument; for fixed L, the δb terms act as a modification of the oscillation wavelength.

An approximate direct limit on δb when $\alpha = \mu$ can be obtained by noting that in atmospheric neutrino data the flux of downward going ν_μ is not depleted whereas that of upward going ν_μ is[10]. Hence, the oscillation arguments in Eqs. (28) and (29) cannot have fully developed for downward neutrinos. Taking $|\delta bL/2| < \pi/2$ with $L \sim 20$ km for downward events leads to the upper bound $|\delta b| < 3 \times 10^{-20}$ GeV for large mixing; upward going events could in principle test $|\delta b|$ as low as 5×10^{-23} GeV. Since the CPT-odd oscillation argument depends on L and the ordinary oscillation argument on L/E, improved direct limits could be obtained by a dedicated study of the energy and zenith angle dependence of the atmospheric neutrino data.

The difference between $P_{\alpha\alpha}$ and $P_{\bar{\alpha}\bar{\alpha}}$

$$P_{\alpha\alpha}(L) - P_{\bar{\alpha}\bar{\alpha}}(L) = -2\sin^2 2\theta \sin \left(\frac{\delta m^2 L}{2E} \right) \sin(\delta bL) , \tag{30}$$

can be used to test for CPT-violation. In a neutrino factory, the ratio of $\bar{\nu}_\mu \to \bar{\nu}_\mu$ to $\nu_\mu \to \nu_\mu$ events will differ from the standard model (or any local quantum field theory model) value if CPT is violated. Fig. 3 shows the event ratios $N(\bar{\nu}_\mu \to \bar{\nu}_\mu)/N(\nu_\mu \to \nu_\mu)$ versus δb for a neutrino factory with 10^{19} stored muons and a 10 kt detector at several values of stored muon energy, assuming $\delta m^2 = 3.5 \times 10^{-3}$ eV^2 and $\sin^2 2\theta = 1.0$, as indicated by the atmospheric neutrino data[10]. The error bars in Fig. 3 are representative statistical uncertainties. The node near $\delta b = 8 \times 10^{-22}$ GeV is a consequence of the fact that $P_{\alpha\alpha} = P_{\bar{\alpha}\bar{\alpha}}$, independent of E, whenever $\delta bL = n\pi$, where n is any integer; the node in Fig. 3 is for $n = 1$. A 3σ CPT violation effect is possible in such an experiment for δb as low as 3×10^{-23} GeV for stored muon energies of 20 GeV. Although matter effects also induce an apparent CPT-violating effect, the dominant oscillation here is $\nu_\mu \to \nu_\tau$, which has no matter corrections in the two–neutrino limit; in any event, the matter effect is in general small for distances much shorter than the Earth's radius.

We have also checked the observability of CPT violation at other distances, assuming the same neutrino factory parameters used above. For $L = 250$ km, the δbL oscillation argument in Eq. (30) has not fully developed and the ratio of $\bar{\nu}$ to ν events is still relatively close to the standard model value. For $L = 2900$ km, a δb as low as 10^{-23} GeV may be observable at the 3σ level. However, longer distances may also have matter effects that simulate CPT violation.

4 Summary

At Long Baseline Experiments and Neutrino Factories true signatures of oscillations (dips) can be established and decay like scenarios can be excluded with confidence. Furthermore these facilities can test CPT conservation at levels better than 10^{-23} GeV.

Acknowledgments
I thank Vernon Barger, Boris Kayser, John Learned, Eligio Lisi, Paolo Lipari, Maurizio Lusignoli, Tom Weiler and Kerry Whisnant for extensive discussions and collaboration. This work was supported in part by U.S.D.O.E under grant DE-FG 03-94ER40833.

References

1. J.G. Learned, S. Pakvasa, and T. J. Weiler, *Phys. Lett.* **B209**, 365 (1988); K. Hidaka, M. Honda, and S. Midorikawa *Phys. Rev. Lett*, **1537** (1988).

2. M.C. Gonzalez-Garcia et al; *Phys. Rev. Lett.* **82**, 3202 (1999).

3. R. Foot, C.N. Leung, and O. Yasuda, *Phys. Lett.* **B443**, 185 (1998).

4. V. Barger et al, *Phys. Rev. Lett.* **82**, 2640 (1999).

5. P. Lipari and M. Lusignoli, *Phys. Rev.* **D60**, 013003 (1999); G.L. Fogli, E. Lisi and A. Marrone, *Phys. Rev.* **D59**, 117 303 (1999); hep/ph-9904248; S. Choubey and S. Goswami, *Astropart. Phys.* **14**, 67 (2000).

6. S. Pakvasa, hep-ph/0004077.

7. J. Valle, *Phys. Lett.* **B131**, 87 (1983); G. Gelmini and J. Valle, *ibid* **B142**, 181 (1983), A. Choi and A. Santamaria, *Phys. Lett.* **B267**, 504 (1991); A. Joshipura and S. Rindani, *Phys. Rev.* **D4**, 300 (1992).

8. V. Barger, W-Y. Keung and S. Pakvasa, *Phys. Rev.* **D25**, 907 (1982).

9. V. Barger, J. G. Learned, P. Lipari, M. Lusignoli, S. Pakvasa and T. J. Weiler, *Phys. Lett.* **B462**, 109 (1999).

10. Y. Fukuda et al.(the Super-Kamiokande Collaboration), *Phys. Rev. Lett.* **81**, 1562 (1998); *ibid* **82**, 2644 (1998).

11. V. Barger, N. Deshpande, P. Pal, R.J.N. Phillips and K. Whisnant, *Phys. Rev.* **D43**, 1759 (1991); E. Akhmedov, P. Lipari and M. Lusignoli, *Phys. Lett.* **B300**, 128 (1993).

12. J. Pantaleone, *Phys. Rev.* **D49**, 2152 (1994).

13. Q. Liu and A. Smirnov, *Nucl. Phys.* **B524**, 505 (1998); P. Lipari and M. Lusignoli, *Phys. Rev.* **D58**, 073005 (1998).

14. Y. Fukuda et al.(the Super-Kamiokande Collaboration), *Phys. Rev. Lett.* **85**, 3999 (2000).

15. F. Vissani and A. Smirnov, *Phys. Lett.* **B432**, 376 (1998); J. Learned, S. Pakvasa, and J. Stone, *Phys. Lett.* **B435**, 131 (1998); L. Hall and H. Murayama, *Phys. Lett.* **B436**, 323 (1998).

16. KEK-PS E362, INS-924 report (1992).

17. MINOS Collaboration, NuMI-L-375 report (1998).

18. ICANOE, F. Emeudo et al., INFN-AE-99-07; OPERA,CERN/SPSC 2000-028.

19. MONOLITH, M. Aglietta et al., CERN/SPSC 98-20.

20. E. Lisi, A. Marrone and D. Montanino, *Phys. Rev. Lett*, **85**, 1166 (2000).

21. R. Alicki and K. Lendi, *Quantum Dynamical Semigroups and Applications*, Lect. Notes Phys. **286** (Springer-Verlag, Berlin, 1987).

22. S.B. Giddings and A. Strominger, *Nucl. Phys.* **B307**, 854 (1988); W. H. Zurek. *Physics Today* **44**, No. 10, p.36 (1991); G. Amelino-Camelia, J. Ellis, N.E. Mavromatos, and D.V. Nanopoulos, *Int. J. Mod. Phys.* **A12**, 607 (1997); L.J. Garay, *Int. J. Mod. Phys.* **A14**, 4079 (1999).

23. J. Ellis, J.S. Hagelin, D.V. Nanopoulos, and M. Srednicki, *Nucl. Phys.* **B241**, 381 (1984).

24. F. Benatti and R. Floreanini, *Phys. Lett.* **B468**, 287 (1999), and references therein.

25. Decoherence as an explanation for atmospheric neutrinos was also suggested in: Y. Grossman and M. Worah, hep-ph/9867511; C.H. Chang et al., *Phys. Rev.* **D60**,033006 (1999).

26. I. Antoniadis, *Phys. Lett.* **B246**,377 (1990); J.D. Lykken, *Phys. Rev.* **D54**, 3693 (1996); N. Arkani-Hamed, S. Dimopoulos and G. Dvali, *Phys. Lett.* **B429**, 263(1998); I. Antoniadis, N. Arkani-Hamed, S. Dimopoulos and G. Dvali, *Phys. Lett.* **B436**, 263 (1998).

27. K.R. Dienes, E. Dudas and T. Gherghetta, *Nucl. Phys.* **B557**,25(1999); N. Arkani-Hamed et al, hep-ph/9811448.

28. R.N. Mohapatra, S. Nandi and A. Perez-Lorenzana, *Phys. Lett.* **B466**, 115 (1999); R. N. Mohapatra and A. Perez-Lorenzana, hep-ph/9910474; Y. Grossman and M. Neubert, hep-ph/9912408.

29. G. Dvali and A. Yu. Smirnov, *Nucl. Phys.* **B563**,63 (1999).

30. R. Barbieri, P-Creminelli and A. Strumia, *Nucl. Phys.* **B585**, 28 (2000); R. Mohapatra, these proceedings.

31. V. Barger, S. Pakvasa, T. Weiler and K. Whisnant, hep-ph/0005197.

32. N. Cabibbo, *Phys. Lett.* **B72**, 333 (1978); V. Barger, K. Whisnant, R.J.N. Phillips, *Phys. Rev. Lett.* **45**, 2084 (1980); S. Pakvasa, in *Proc. of the XXth International Conference on High Energy Physics*, ed. by L. Durand and L.G. Pondrom, AIP Conf. Proc. No. 68 (AIP, New York, 1981), Vol. 2, p. 1164.

33. D. Colladay and V.A. Kostelecky, *Phys. Rev.* **D55**, 6760 (1997).
34. S. Coleman and S.L. Glashow, *Phys. Rev.* **D59**, 116008 (1999).
35. V. Barger, K. Whisnant, S. Pakvasa, and R.J.N. Phillips, *Phys. Rev.* **D22**, 2718 (1980).
36. L. Wolfenstein, *Phys. Rev.* **D17**, 2369 (1978); S.P. Mikheyev and A. Smirnov, *Yad. Fiz.* **42**, 1441 (1985) [*Sov. J. Nucl. Phys.* **42**, 913 (1986)].

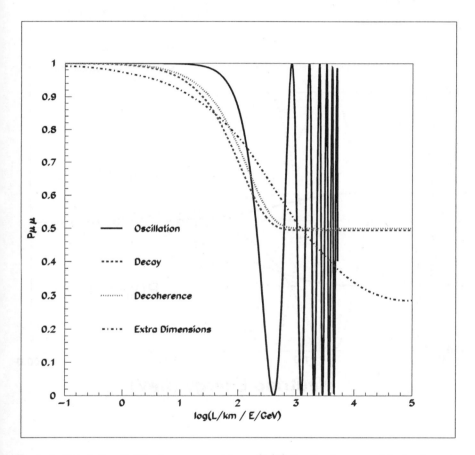

Figure 1. Survival probality for ν_μ versus log_{10} (L/E) for the decay model, decoherence, extra dimensions and oscillation.

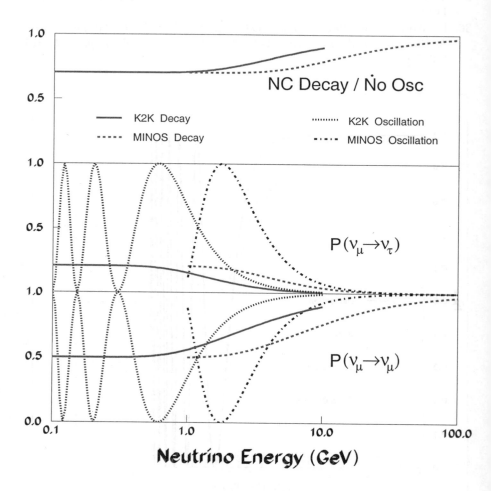

Figure 2. Long-baseline expectations for the K2K and MINOS long-baseline experiments from the decay model and the ν_μ-ν_τ oscillation model. The upper panel gives the neutral current predictions compared to no oscillations (or ν_μ-ν_τ oscillations).

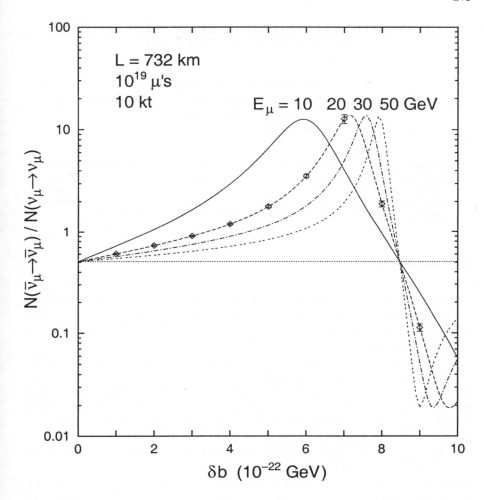

Figure 3. The ratio of $\bar{\nu}_\mu \to \bar{\nu}_\mu$ to $\nu_\mu \to \nu_\mu$ event rates in a 10 kt detector for a neutrino factory with 10^{19} stored muon with energies $E_\mu = 10, 20, 30, 50$ GeV for baseline $L = 732$ km versus the CPT-odd parameter δb with $\theta_m = \theta_b \equiv \theta$ and phase $\eta = 0$. The neutrino mass and mixing parameters are $\delta m^2 = 3.5 \times 10^{-3}$ eV^2 and $\sin^2 2\theta = 1.0$. The dotted line indicates the result for $\delta b = 0$, which is given by the ratio of the $\bar{\nu}$ and ν charge-current cross sections. The error bars are representative statistical uncertainties.

CONVENTIONAL NEUTRINO BEAM EXPERIMENT: JHF - SUPER-KAMIOKANDE

YOSHIHISA OBAYASHI

Kamioka Observatory, ICRR, Univ. of Tokyo, Higashi-Mozumi, Kamioka-cho,
Gifu 506-1205, Japan
E-mail: ooba@icrr.u-tokyo.ac.jp

A long baseline experiment is under preparation to start from the year 2006 with a conventional neutrino beam from JHF and Super-Kamiokande as the far detector. The precision for the ν_μ disappearance channel are $\delta(\Delta m^2) \sim 2 \times 10^{-4} eV^2$ and $\delta(\sin^2 2\theta_{23}) \sim 0.01$ by five year exposure. The sensitivity for the ν_e appearance channel is $\sin^2 2\theta_{13} \sim 6 \times 10^{-3}$ at the atmospheric neutrino best fit point.

1 Introduction

Super-Kamiokande atmospheric neutrino results[1] have presented the evidence that neutrinos have finite mass and oscillate between their flavors. Next generation neutrino experiments are expected to "measure" the oscillation parameter as precisely as possible.

A long baseline neutrino oscillation experiment which will measure oscillation parameters with high precision is now under preparation and expected to start its run from 2006. Neutrino beam will be produced by conventional beam line in "Japan Hadron Facility (JHF)", which will be constructed in JAERI, Tokai-village. As the far detector, the already existing water cherenkov detector, Super-Kamiokande(SK) will be used.

In the three neutrino flavor framework, the oscillation probabilities from ν_μ to other flavors are described as: $P(\nu_\mu \rightarrow \nu_x) = \sin^2 2\theta_{\mu x} \cdot \sin^2(1.27 \cdot \Delta m^2 \cdot L/E)$, $\sin^2 2\theta_{\mu\tau} = \cos^4 \theta_{13} \cdot \sin^2 2\theta_{23}$, $\sin^2 2\theta_{\mu e} = \sin^2 2\theta_{13} \cdot \sin^2 \theta_{23}$. Here, the mass hierarchy is assumed as $\Delta m^2 \equiv \Delta m_{23}^2 \gg \Delta m_{12}^2$. Then, θ_{23} and Δm^2 can be measured by observing the ν_μ disappearance probability $P(\nu_\mu \rightarrow \nu_x) = P(\nu_\mu \rightarrow \nu_\tau) + P(\nu_\mu \rightarrow \nu_e)$ as a function of neutrino energy, and, θ_{13} can be measured by observing the $\nu_\mu \rightarrow \nu_e$ appearance probability $P(\nu_\mu \rightarrow \nu_e)$.

The precise measurement of the mass difference Δm_{23}^2 and the mixing angle θ_{23}, θ_{13} is also an important step to the further physics like the CP violation measurement in the lepton sector, or measurement of matter effects in the earth.

Figure 1. Geometrical overview of the JHF-SK neutrino experiment. The neutrino beam will be produced in JAERI, Tokai-village, and Super-Kamiokande in Kamioka-town will be used as the far detector. The baseline is 295 km.

2 Neutrino Beam

The neutrino beam will be produced in the Japan Atomic Energy Research Institute (JAERI) with a 50GeV proton driver and conventional neutrino beam line where pions decay into muons and neutrinos. JHF will inject 1×10^{21} protons on target per year(130day run) for neutrino production.

Several types of the configurations of beam line setup are considered. Our present strategy is to start from a wide band beam(WBB) and pin down the mass difference in the first one year run. Then, we will switch to narrow band beam. After the five-years of narrow band beam exposure, we would measure the mixing precisely. Two options for the narrow band beam are considered. One is a conventional narrow band beam line (NBB) which selects pion momenta by dipole magnet. Another option is the challenging "off axis" beam line (OAB) which came up in the discussion of BNL E889[3]. The beam line itself is almost the same as WBB, but detectors are put at intentionally misaligned positions. Hence, we can expect a more intense narrow band beam than conventional NBB. Expected beam spectra for the three beam configurations are shown in Figure 3 and expected numbers of neutrino interactions are summarized in Table 1.

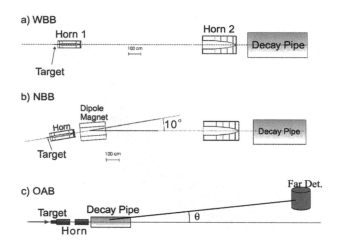

Figure 2. Schematic views of possible options for beam line configuration.

Table 1. Expected fluxes for three possible configurations of neutrino beam line.

	$N_{int.}$@SK(22.5kt)	ν_e contamination (%)
WBB	4200	0.8
NBB w/dipole	830	0.3
2° Off Axis NBB	2200	0.2

3 Physics Sensitivity

3.1 ν_μ disappearance search

In the case that $\sin^2 2\theta_{13} \ll \sin^2 2\theta_{23}$, the ν_μ disappearance probability can be described as $P(\nu_\mu \not\to \nu_\mu) = \sin^2 2\theta_{23} \cdot \sin^2(1.27\Delta m^2 L/E)$ hence the measurement of θ_{23} and Δm^2 can be performed without the uncertainty from the value of θ_{13}. In the energy region shown in Figure 3, the main interaction mode is the quasi-elastic interaction. Therefore, we will select only events with a muon track ("single-ring μ-like", developed in the atmospheric neutrino analysis in SK). Then, a parameter fit with mass difference and mixing on the reconstructed neutrino energy distribution will be applied. An example of a fit result for a parameter set is shown in Figure 5. Estimated resolutions are $\delta(\Delta m^2) \sim 2 \times 10^{-4} eV^2$ even by a 1 year exposure of any beam configuration and $\delta(\sin^2 2\theta_{23}) \sim 1\%$ by a 5 year exposure of NBB. Variation of resolutions

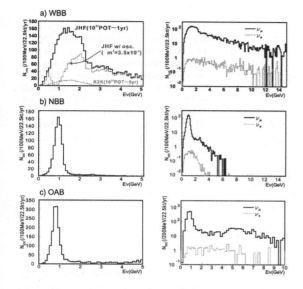

Figure 3 Expected event rates by the function of the neutrino energy for one year exposure of three possible beam configurations. Beam ν_e contaminations are shown in right figures as well.

for possible combinations of true parameters are shown in Figure 6.

3.2 ν_e appearance search

In the measurement of the mixing θ_{13} via $\nu_\mu \to \nu_e$ appearance, the reduction of background mainly from neutral current π^0 production and estimation precision of the backgrounds are key techniques to increase its sensitivity.

The first estimate for ν_e appearance search is done with the ν_e C.C. selection used in the atmospheric neutrino analysis in SK (single-ring e-like). The reduction rate and survival efficiencies are shown in Table 2. We have developed a more powerful π^0 rejection algorithm and applied it in the estimation. The main feature of this advanced analysis is the program is forced to find a second track(ring) when applied to the single ring e-like sample. It then reconstructs the quantities shown in Figure 7 and π^0-like events are removed using these quantities. Table 2 shows the selection efficiencies and BG contamination rates for the three beam configurations. Expected neutrino energy distributions which reconstructed after the π^0 rejection are shown in Figure 8. Sensitive oscillation parameter regions for three beam configurations are shown in Figure 9. The OAB configuration has the best sensitivity

220

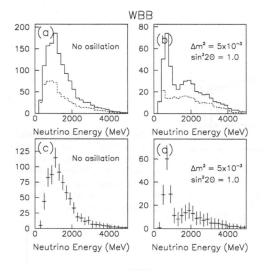

WBB

(a) No osillation

(b) $\Delta m^2 = 5\times10^{-3}$
$\sin^2 2\Theta = 1.0$

Neutrino Energy (MeV)

(c) No osillation

(d) $\Delta m^2 = 5\times10^{-3}$
$\sin^2 2\Theta = 1.0$

Neutrino Energy (MeV)

Figure 4   The expected energy spectra of single ring μ-like events. Top figures are without subtracting the non-QE contributions, and, bottom figures are with subtracting. Left figures are for no oscillation and right figures are for oscillation with parameters: $(\Delta m^2_{23}, sin^2 2\theta_{23}) = (5 \times 10^{-3} eV^2, 1.0)$. Non-QE contributions are shown by dashed line in figure (a) and (b).

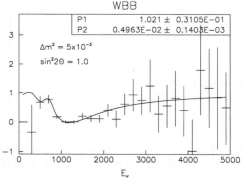

WBB

P1	1.021 ± 0.3105E−01
P2	0.4963E−02 ± 0.140 E−03

$\Delta m^2 = 5\times10^{-3}$
$\sin^2 2\Theta = 1.0$

E_ν

Figure 5   Expected ratio of the energy spectra between the observed events with oscillations and the expected one for no oscillation. Fit results by the function: $P = \sin^2 2\theta_{23} \cdot \sin^2(\Delta m^2_{23} \cdot 295km/E)$ are shown as well.

for this ν_e appearance search and the sensitive region is $\sin^2 2\theta_{\mu e} > 3 \times 10^{-3}$ at $\Delta m^2 \sim 3 \times 10^{-3} eV^2$ mass region. When we use $\sin^2 2\theta_{23} = 1.0$, which is from the atmospheric neutrino best fit value, $\sin^2 2\theta_{\mu e} = 3 \times 10^{-3}$ can be converted to $\sin^2 2\theta_{13} = 6 \times 10^{-3}$.

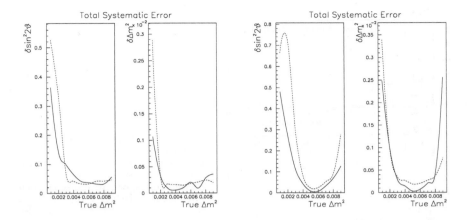

Figure 6. The variation of total statistical errors (dashed line) and systematic uncertainties (solid line) as a function of true (input) Δm^2. Left two figures are for 1 year exposure of WBB, and right figures are for 1 year exposure of NBB.

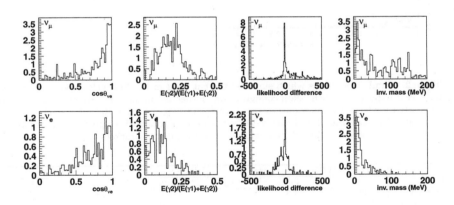

Figure 7. Distributions of the quantities used in the π^0 reduction for WBB upper and lower histograms correspond to the events from ν_μ interactions (mainly from π^0 production) and ν_e interactions, respectively.

222

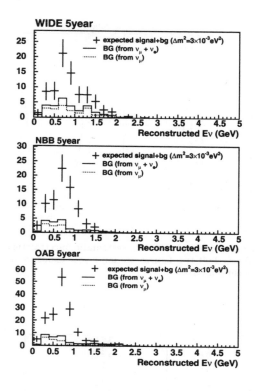

Figure 8   Expected reconstructed neutrino energy distribution for 5 year exposure of the three beam configurations. The cross, line, and dashed histogram correspond to Expected oscillated signal($\sin^2 2\theta_{\mu e} = 0.05$, $\Delta m_{23}^2 = 3 \times 10^{-3} eV^2$) + BG, All BG, and BG from ν_μ interaction, respectively.

Table 2. Number of events which survived the cut used in the LOI (1R e-like) and π^0 cut for 1 year exposure of each beam configuration. Expected number of signal interactions are calculated for the oscillation parameters $(\Delta m_{23}^2, \sin^2 2\theta_{\mu e}) = (3 \times 10^{-3} eV^2, 0.05)$. Reduction rates for background and survival efficiencies for signal with respect to the number of events reconstructed in the fiducial volume are shown in the brackets.

	WBB		NBB		OAB	
	ν_μ	ν_e	ν_μ	ν_e	ν_μ	ν_e
Fid.Vol.	3651	53.3	740	23.0	1801	45.5
1R e-like	92(2.5%)	29.5(55.3%)	13.2(1.8%)	16.2(70.4%)	37.4(2.1%)	32.1(70.5%)
π^0 cut	4.0(0.1%)	9.8(18.4%)	1.8(0.2%)	11.6(50.4%)	3.8(0.2%)	24.3(53.4%)

4 Summary

A next generation long baseline neutrino experiment to measure some oscillation parameters is expected to start from the year 2006 in Japan. Thanks

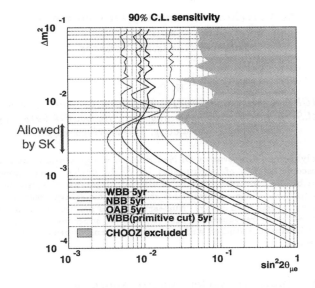

Figure 9. Contour plots of sensitivity for ν_e appearance search.

to the high intensity of the neutrino beam and large detector mass, estimated resolutions of the measurement are typically $\delta(\Delta m^2) \sim 2 \times 10^{-4} eV^2$, $\delta(\sin^2 2\theta_{23}) \sim 1\%$ and 90% C.L. sensitivity for $\sin^2 2\theta_{13} \sim 6 \times 10^{-3}$ at the parameter region of atmospheric neutrino results.

References

1. Y. Fukuda *et al.*, Phys. Rev. Lett. **81**, 1562 (1998)
2. "Letter of Intent: A Long Baseline Neutrino Oscillation Experiment using the JHF 50 GeV Proton-Synchrotron and the Super-Kamiokande Detector", Feb. 3, 2000, http://neutrino.kek.jp/jhfnu
3. BNL E889 proposal, 1995

MOON(Mo Observatory Of Neutrinos)
and
Intense Neutrino Sources

H. Ejiri
RCNP, Osaka University, Ibaraki, Osaka, 567-0047, Japan
JASRI, SPring-8, Mikazuki-cho, Sayo-gun, Hyogo, 675-5918, Japan
ejiri@rcnp.osaka-u.ac.jp, ejiri@spring8.or.jp

MOON(Molybdenum Observatory Of Neutrinos) is a high sensitive detector used for spectroscopic studies of neutrinos(ν) in ^{100}Mo nuclear micro-laboratories. MOON can be used for spectroscopic studies of Majorana ν masses by double beta decays($\beta\beta$) of ^{100}Mo and real-time studies of low energy solar ν's and supernova ν's by inverse beta decays. Nuclear responses are crucial for ν studies in nuclei. Nuclear weak responses for $\beta\beta - \nu$, solar-ν, supernova-ν, and for ν's of astrophysics interest are studied well by means of intense ν beams combined with large ν detectors.

1 Neutrino studies in nuclear micro-laboratories

Neutrinos(ν) and weak interactions are windows for new physics beyond the standard electroweak theory of $SU(2)_L \times U(1)$. Fundamental properties of ν's and weak interactions are studied in nuclear micro-laboratories [1]. Actually Majorana ν masses, Mojoron-ν and SUSY-ν couplings, right-handed ν's and weak bosons, ν oscillations and others are well studied by investigating double beta($\beta\beta$) decays in nuclei, solar- and supernova-ν interactions(inverse β decays) in nuclei, and other low energy ν-nuclear processes of astroparticle physics interest [2][3][4].

Neutrino mass is a key issue of current neutrino(ν) physics. Recent results with atmospheric, solar and accelerator ν's strongly suggest ν oscillations due to non-zero ν-mass differences and flavor mixings.

Double beta decay may be the only probe presently able to access such small ν masses. Actually, observation of neutrino-less double beta decay ($0\nu\beta\beta$) would identify a Majorana-type electron ν with a non-zero *effective* mass $< m_\nu >$ [2]. The $0\nu\beta\beta$ process is, in fact, sensitive not only to the ν mass ($< m_\nu >$) but also to a right-handed weak current and other terms beyond the Standard Model(SM) [2][3].

Low-energy solar-ν studies are crucial for investigating fundamental properties of electron ν's. So far, they have been studied by non-real-time and inclusive measurements that do not measure the neutrino energies and thus do not identify the ν sources in the sun. Real-time spectroscopic studies of low energy solar-ν are important for studies of the solar-ν problems.

2 Neutrino studies in ^{100}Mo

Recently it has been shown that measurements of two correlated β rays from ^{100}Mo make it possible to perform both spectroscopic studies of $0\nu\beta\beta$ with a sensitivity of the order of $< m_\nu > \sim 0.03$ eV, and real-time exclusive studies of low energy solar ν by inverse β decay[5]. The unique features are as follows.

1)The β_1 and β_2 with the large energy sum of $E_1 + E_2$ are measured in coincidence for the $0\nu\beta\beta$ studies, while the inverse β-decay induced by the solar ν and the successive β-decay are measured sequentially in an adequate time window for the low energy solar-ν studies. The isotope ^{100}Mo is just the one that satisfies the conditions for the $\beta\beta - \nu$ and solar-ν studies.

2)The large Q value of $Q_{\beta\beta}=3.034$ MeV gives a large phase-space factor $G^{0\nu}$ to enhance the $0\nu\beta\beta$ rate and a large energy sum of $E_1 + E_2 = Q_{\beta\beta}$ to place the $0\nu\beta\beta$ energy signal well above most BG except ^{208}Tl and ^{214}Bi. The energy and angular correlations for the two β-rays can be used to identify the ν-mass term.

3)The low threshold energy of 0.168 MeV for the solar-ν absorption allows observation of low energy sources such as pp and ^{7}Be. The GT strength to the 1^+ ground state of ^{100}Tc is measured to be $(g_A/g_V)^2 B(GT)=0.52\pm0.06$ by both charge-exchange reaction and electron capture[6][7]. Capture rates are large even for low energy solar ν's, as shown in Table 1. The solar-ν sources are identified by measuring the inverse-β energies. Only the ^{100}Tc ground state can absorb ^{7}Be-ν and pp-ν. Therefore the intensity ratio of the pp-ν and the ^{7}Be-ν is independent of the B(GT).

4)The measurement of two β-rays (charged particles) enables one to localize in space and in time the decay-vertex points for both the $0\nu\beta\beta$ and solar-ν studies. Radiations associated with BG are also measured. The tightly localized β-β event in space and time windows, together with relevant β and γ measurements, are key points for selecting $0\nu\beta\beta$ and solar-ν signals and for reducing correlated and accidental BG by factors $10^{-5} \sim 10^{-6}$ as in the case of ELEGANT[8].

3 MOON: Molybdenum Observatory Of Neutrinos

The $0\nu\beta\beta$ transition rate $R_{0\nu}$ for $< m_\nu >$ is given by

$$R_{0\nu} = G^{0\nu}(M^{0\nu})^2| < m_\nu > |^2, \tag{1}$$

Table 1: Solar-ν absorption rates R_ν for ^{100}Mo.[5]

Source	$E_\nu^{(max)}$(MeV)	$E_\beta^{(max)}$(MeV)	R_ν/SNU^a
pp	0.42	0.25	639 ± 85
pep	1.44	1.27	13 ± 2
^{7}Be	0.86	0.69	206 ± 35
^{8}B	~ 15	~ 14.2	$27(23)^b \pm 4$
^{13}N	1.20	1.03	22 ± 3
^{15}O	1.74	1.57	32 ± 4

$E_\nu^{(max)}$, $E_\beta^{(max)}$ are the maximum ν energy and maximum β-ray energy.
a) Standard-solar-model(SSM) capture rates based on BP98[4] with errors from those of $B(GT)$.
b) Rate for the states below the effective neutron threshold energy.

where $G^{0\nu}$ is the phase space factor and $M^{0\nu}$ is the matrix element, both relatively large for ^{100}Mo.

The $g_{7/2}^\nu - g_{9/2}^\pi$ shell-model structure of ^{100}Mo -^{100}Tc leads to the large measured $2\nu\beta\beta$ decay rate[8], and the large calculated value for the $0\nu\beta\beta$ decay rate[2].

The $0\nu\beta\beta$ events are identified by setting the appropriate energy window and the prompt time window for the $\beta\beta$ coincidence signals. The rate in units of 10^{-36}/sec is given as $R_{0\nu} = 6.6 \times 10^4 | < m > |^2/(\text{eV})^2$ by RQRPA[3]. The uncertainty in calculation of the nuclear matrix element is considered to be of order 50 %.

For solar ν detection, the inverse β-decay induced by the solar-ν absorption is followed by β-decay with a mean life $\tau = 23$ sec. Thus a time window can be set as $\Delta T = 30$ sec($10^{-6}y$) from $t_1 = 1$ sec to $t_2 = 31$ sec. The starting time of 1 sec is long enough to reject most correlated BG such as the $2\nu\beta\beta$, β-rays followed by conversion electrons, scatterings of single β-rays, etc. The stopping time of 31 sec is short enough to limit the accidental coincidence BG. The accidental rate is further reduced by effectively subdividing the detector into K unit cells by means of position readout.

The lower limit (sensitivity) on $< m_\nu >$ can be obtained by requiring that the number of $0\nu\beta\beta$ events has to exceed the statistical fluctuation of the BG events. The sensitivity of the order of $< m_\nu > \sim 0.03$ eV can be achieved for three year measurement by means of a realistic detector with a few tons of ^{100}Mo and RI contents of the order of 0.1ppt($b \sim 10^{-3}$Bq/ton).

Sensitivity for the solar ν is obtained similarly as in case of the $0\nu\beta\beta$. It is of the order of ~ 100 SNU for one year measurement by using the same detector with $K \sim 10^9$. In fact the $2\nu\beta\beta$ rate and the BG rate from RI at 0.1ppt($b \sim 10^{-3}$Bq/ton) are larger than the solar-ν rate by factors $\sim 10^7$ and $\sim 10^5$, respectively. The fine localization in time($\Delta T = 10^{-6}y$) and in space($1/K = 10^{-9}$), which is possible with the present two-β spectroscopy, is crucial for reducing BG rates in realistic detectors.

One possible detector is a super-module of ~ 1 ton of ^{100}Mo (~ 10 tons of Mo in case of natural Mo) purified to 10^{-3} Bq/ton for ^{238}U and ^{232}Th or less. This purity level has been achieved for Ni and other materials for the Sudbury Neutrino Observatory[9]. The super-module with a fiducial volume of (x, y, z)=(2.5m,2.5m,1~5m, depending on enriched ^{100}Mo or natural Mo) is composed of 400~2000 modules with (x, y, z) =(2.5m,2.5m,0.25cm). The Mo foils with thickness of 0.05~0.03 g/cm^2 are interleaved between the modules. Light outputs from each scintillator module are collected by WLS(wave length shifter) fibers. One may use only syntillation fiber arrays streched to x and y directions in place of scintillator/WLS ensembles. Use of enriched ^{100}Mo isotopes with 85 % enrichment is very effective for reducing the detector volume and for getting the large S/N ratio.

The detector can be used also for supernova-ν studies and other rare nuclear processes, and for other isotopes. Another option is a liquid scintillator in place of the solid one, keeping similar configurations of the WLS readout[10]. The energy and spatial resolution are nearly the same. Then ^{150}Nd with the large $Q_{\beta\beta}$ may be used either in solid or solution in the liquid scintillator for $0\nu\beta\beta$. Of particular interest is ^{136}Xe because liquid Xe is a scintillator.

4 Nuclear responses for neutrinos

Nuclei as quantum systems of nucleons are used as micro-laboratories for studying particle physics beyond the standard electro-weak theory of $SU(2)_L \times U(1)$ and new aspects of astronuclear processes in the sun and stars[1], as discussed in sections 1 and 2. The Majorana ν masses, the right-handed weak interactions and the others are effectively studied by $\beta\beta$ decays in nuclei. Solar and supernova ν's provide important information on ν oscillations, ν masses and stellar evolution mechanisms. They are studied by using nuclear inverse β decays by the charged-current interaction and nuclear inelastic scatterings by the neutral-current interaction. Nuclear weak processes are important also for astronuclear reactions and nuclear synthesis. Then nuclear responses for the $\beta\beta - \nu$, the solar and supernova ν's and the astronuclear weak processes are crucial for the studies of the neutrinos and the weak processes in nuclei[1][11].

5   Nuclear responses by electromagnetic, strong and weak probes

The weak interactions in nuclei involve spin and isospin dependent terms, and accordingly the nuclear weak responses are greatly modified in nuclei by nuclear spin isospin medium effects which reflect the nuclear spin isospin structures. Therefore the nuclear spin isospin structures play essential roles for studies of the nuclear weak processes.

The spin isospin operator is in general expressed as

$$T_{TSLJ} = g^{\alpha}_{TSLJ} \tau^{T} [r^{L} Y_{L} \times \sigma^{S}]_{J}, \tag{2}$$

where T, S, L, and J are the isospin, spin, orbital angular momentum and total angular momentum, respectively, and Y_{L} is the spherical harmonics. The coupling constant is given by g^{α}_{TSLJ} with α standing for the weak, electromagnetic or strong interaction.

The vector and axial-vector weak operators are given, respectively, by the isospin and spin-isospin operators of T_{10LJ} and T_{11LJ} with $T = T_{\pm}$ for the charged-current interaction and $T = T_{3}$ for the neutral-current one. The electric and magnetic operators are written similarly by the spin isospin operators with the isospin $T = T_{3}$.

The nuclear weak responses investigated by nuclear β decays are limited to the charged-current ground state transitions.

Hadronic nuclear reactions have been extensively used to study nuclear spin isospin responses in a wide excitation region, charge-exchange spin-flip and non spin-flip reactions for the charged-current spin-isospin and isospin responses and inelastic scatterings for the neutral-current responses [1].

The hadronic processes, however, involve necessarily strong interactions mediated by various kinds of mesons, and accordingly the spin isospin interaction is not just the simple one of the central spin isospin interaction, but include rather complicated terms. In some cases tensor interactions play important roles. The interaction strength itself is strong and thus the reaction proceeds by multi-step processes as well as the direct single step process. Projectile particles and emitted particles in nuclear reactions are distorted by the strong nuclear potential. Therefore it is not straightforward to deduce the spin isospin responses from hadronic reactions. In fact, they are used only for limited cases of the strong spin-isospin excitations with $TSLJ=1101$ modes in the $T_{\pm}$ channel.

Electromagnetic probes such as real photons, real electrons and virtual photons associated with high-Z projectiles are used to study nuclear spin isospin structures and responses. These electromagnetic probes, however, include isoscalar currents and orbital currents as well. In general it is not easy to separate experimentally these contributions from isovector spin currents relevant to the weak interaction. The isospin spin responses studied by the electromagnetic probes are only for the T_3 channel.

The neutrino probe is an excellent and ideal probe for studying the sin isospin structures and the spin isospin responses of the nuclear and astroparticle physics interests. It can probe directly for the spin isospin structures and the spin isospin weak responses through the charged- and neutral-current interactions. In contrast to hadron projectiles, ν is free from the nuclear distortion and the multi-step process since the interaction is very weak.

6 Intense neutrino sources for neutrino response studies

Neutrinos involved in neutrino and weak processes in nuclei are mostly low energy ν's up to 50 MeV. Intense ν sources in the low energy region of $1\sim50$ MeV are obtained from $\pi - \mu$ decays, and π beams are produced by medium energy proton beams. The reaction rate for low energy ν's is necessarily extremely small. Thus high intensity proton accelerators combined with large efficiency ν detectors are of vital importance.

One possible accelerator/detector ensemble is obviously SNS/ORLaND at ORNL [10]. SNS with 10^{16}/sec 1GeV protons will give 7 10^{14} low energy ν's per sec. A high intensity accelerator complex at JAERI/Tokai is also of potential interest for low energy ν sources. The 3 GeV PS with 2 10^{15} protons can be used to provide intense low energy ν's with 3 10^{14}/sec. Then one needs to build a large ν detector for general use. A detector like MOON is useful for this type of experiments.

Thus SNS/ORLaND and JAERI/ν-detector are expected to play leading roles as low energy ν factories to study neutrino nuclear physics of nuclear, particle and astrophysics interests.

Some of interesting subjects to be studied by means of intense ν sources are as follows.

1. Spin isospin responses for the charged-and neutral-current isospin $(TSLJ = 10LL)$ and spin-isospin$(TSLJ = 11LJ)$ modes with $L = 0, 1, 2$ in a wide excitation region. They elucidate nuclear spin isospin structures of the nuclear and astroparticle physics interests, including spin isospin core polarizations and spin isospin giant resonances with $S=1$ and 0 and $L=0,1,2$ in both $T = T_\pm$ and T_3 channels.

2. Spin-isospin responses for ^{71}Ga, ^{100}Mo, ^{176}Yb and other nuclei used for solar neutrino studies. The responses for all excited states in these nuclei are estimated from hadronic charge-exchange reactions [6][12]. The neutrino probe

gives ambiguously pure spin isospin responses. They are quite important for quantitative studies of ν oscillations.

3. Spin isospin responses with $L=0$ and 1 for ^{16}O, ^{100}Mo, ^{208}Pb and others, which are used for supernova ν studies. The spin isospin responses in a wide excitation region of the $T_\pm = 1$ and T_3 channels are essential for quantitative studies of ν oscillations, ν masses, stellar evolution and synthesis processes and others of the astroparticle physics interests.

4. Nuclear spin isospin strength distributions with $L=0\sim 5$ in the T_- channel for ^{100}Mo, ^{116}Cd, ^{136}Xe, and others nuclei used for studies of double beta decays. The responses are important for studying the properties of the neutrinos and the weak interactions by 0ν $beta\beta$.

5. Comparison of neutrino weak processes with corresponding hadronic and electromagnetic processes to investigate reaction mechanisms involved in the hadronic and electromagnetic processes and the isoscalar and orbital magnetic currents.

6. Nuclear weak responses associated with SN evolutions, nuclear synthesis and other nuclear weak processes of astrophysics interests.

The author thanks prof. F.Avignone and Prof.R.G.H.Robertson for valuable discussions, and Prof. Y.Kuno for encouragement for presenting this work in LFV workshop.

References

1. H .Ejiri, Int. J. Mod. Phys. E6 No 1 (1997) 1;
   H. Ejiri, Phys. Rep. C338 (2000) 265.
   H.Ejiri, Nucl.Phys.B Suppl.Conf. Proc. ν-2000. to be published.
2. W. C. Haxton and G. J. Stephenson Jr, Prog. Part. Nucl. Phys. 12 (1984) 409.
   M. Doi et al., Prog. Theor. Phys. 83 (Suppl.)(1985) 1;
3. A. Faessler and F. Simcovic, J. Phys. G 24 (1998) 2139
4. J. N. Bahcall and M. Pinsonneault, Rev. Mod. Phys. 64 (1992) 885, and 6 7 (1995) 781.
   J. N. Bahcall et al., Phys. Lett. B433 (1998) 1.
5. H. Ejiri, J. Engel, R. Hazama, P. Krastev, N. Kudomi, and R.G.H. Robertson, Phys. Rev. Lett.85 (2000) 2917, nuclexp/9911008 v2, 23 Nov 1999.
6. H. Akimune, et al., Phys. Lett. B394 (1997) 23.
   H.Ejiri, et al., Phys. Lett,B433 (1998) 257.
7. A. García et al., Phys. Rev. C47 (1993) 2910.
8. H. Ejiri et al., Phys. Lett. B258 (1991) 17.
   H. Ejiri et al., Nucl. Phys. A611 (1996) 85.

H. Ejiri, Nucl. Phys.**A.577** (1994) 399c.

J. Phys. Sos. Japan Lett. **65** (1996) 7.

9. R.G.H.Robertson, Prog. Part. Nucl. Phys. **40** (1998) 113.

10. F. Avignone III et al., ORLaND Coll, ORLaND proposal (1999).Nucl.Phys. **B** Proc. Suppl. tp be published.

11. H. Ejiri and J.I. Fujita, Phys. Rep. **38 C** (1978) 85.

    H. Ejiri, Nucl. Phys.**A.577** (1994) 399c; J. Phys. Sos. Japan Lett. **65** (1996) 7.

12. M.Fujiwara, et al.,Phys. Rev. Lett. **85**(2000) 4442.

NEXT GENERATION WATER CHERENKOV DETECTOR AT KAMIOKA

K. NAKAMURA

High Energy Accelerator Research Organization (KEK), Tsukuba, Ibaraki 305-0801, Japan
E-mail: kenzo.nakamura@kek.jp

M. SHIOZAWA

Kamioka Observatory, ICRR, University of Tokyo, Higashi-Mozumi, Kamioka-cho, Yoshiki-gun, Gifu 506-1205, Japan
E-mail: masato@icrkm4.icrr.u-tokyo.ac.jp

Hyper-Kamiokande, a very large water Cherenkov detector with mass of ~ 1 Mton, is discussed for a next generation nucleon decay experiment at Kamioka. Other physics possibilities with this detector are also mentioned. The sensitivities of such a large water Cherenkov detector are studied for the $p \to e^+\pi^0$ decay mode, assuming 40%, 10%, and 4.4% photocathode coverages, with the same or similar selection criteria as the Super-Kamiokande analysis and with a tighter total momentum cut.

1 Introduction

Evidence of finite neutrino mass found by the Super-Kamiokande (hereafter abbreviated as SK) Collaboration through atmospheric-neutrino observations [1] provided a major breakthrough in our understanding of nature. Extremely small but finite neutrino mass indicates physics beyond the Standard Model. Most of unified gauge theories can predict relevant neutrino mass by introducing new physics at much higher energy scale than the electroweak scale. These theories also predict nucleon instability. Consequently, improved search for nucleon decay has become increasingly more important.

At present, SK, the world-largest nucleon-decay experiment, sets lower limits of nucleon partial lifetime for the two most interesting channels as follows [2]:

$$\tau_p/B(p \to e^+\pi^0) > 4.4 \times 10^{33} \text{ yr}$$

$$\tau_p/B(p \to \bar{\nu}K^+) > 1.9 \times 10^{33} \text{ yr},$$

where B means the branching ratio. These limits were obtained with an exposure of 70 kton·year. In less than 10 years, SK will reach a $\tau_p/B(p \to e^+\pi^0)$ limit of 10^{34} years and the $\tau_p/B(p \to \bar{\nu}K^+)$ limit of $\sim 4 \times 10^{33}$ years.

Among unified gauge theories, supersymmetric (SUSY) grand unified theories are particularly interesting because they can solve the mass hierarchy problem and they successfully predict $\sin^2\theta_W$. The unification scale is $\sim 10^{16}$ GeV and a gauge-boson-mediated decay mode $p \to e^+\pi^0$ has a predicted lifetime of $\sim 10^{36}$ years.[a] In SUSY grand unified theories, nucleon decay also occurs through dimension 5 operators; the dominant decay modes involve K mesons. The present SK limit on the $p \to \bar{\nu}K^+$ decay mode already excludes the minimal SUSY SU(5). However, the predicted lifetime limit of this decay mode is highly model-dependent, and it is important to reach longer lifetime limits of the decay modes involving K mesons.

2 Next-Generation Water-Cherenkov Detector

To reach the $p \to e^+\pi^0$ partial lifetime limit of 10^{36} years within a reasonable observation time, a detector with 100 times the SK's mass (50,000 tons) would be needed even if this decay mode were background-free up to such a long lifetime limit. This, however, seems too big a step forward. A reasonable size of the next generation of nucleon decay experiment may be $10 \sim 20$ times the SK's mass, namely, $0.5 \sim 1$ Mtons. Assuming $10 \times$ SK mass and 10 years of observation, a very crude estimation of expected partial lifetime limits is

$$\tau_p/B(p \to e^+\pi^0) > 10^{35} \ \text{yr}$$

$$\tau_p/B(p \to \bar{\nu}K^+) > 10^{34} \ \text{yr},$$

where the $p \to e^+\pi^0$ decay mode is assumed to be background free. (As discussed later in this paper, however, this decay mode is not completely backgroud free in the domain of 10^{35} years of partial lifetime.) For the $p \to \bar{\nu}K^+$ decay mode, background dominance is assumed.

For such a very big detector, we envisage to employ water-Cherenkov technique because water is the cheapest detector material and one order of magnitude extension of the well-proven SK will not cause any serious difficulties both in construction and in operation. We call the next-generation water Cherenkov detector **Hyper-Kamiokande**,[b] explicitly assuming that it is constructed at the Kamioka Observatory. [4,5]

---

[a] In this workshop, Marciano pointed out [3] that the central prediction of SUSY SU(5) on the $\tau_p/B(p \to e^+\pi^0)$ is 10^{35} years rather than 10^{36} years.

[b] In a meeting held in Japan, [5] there was a comment that *super-* is a prefix of Latin origin and *hyper-* is a prefix of Greek origin, both having the same meaning, and thus it is misusage to call a larger detector than Super-Kamiokande as Hyper-Kamiokande. However, there is a few cases in which *hyper-* is used to indicate "beyond *super-*." For example, while *supersonic* (and also *ultrasonic*) indicates speeds from one to five times the speed of sound in air, *hypersonic* does indicate speed five times or more that of sound in air.

Hyper-Kamiokande may be able to address other interesting physics in addition to nucleon decay.

Presently, SK is used as a far detector for the K2K (KEK-to-Kamioka) long-baseline neutrino oscillation experiment. It will also be used as a far detector in a future long-baseline neutrino oscillation experiment with a high-intensity ($\sim$ 1 MW beam power) 50-GeV proton synchrotron[c] to be built at JAERI (Japan Atomic Energy Research Institute) in Tokai-mura as a neutrino source. [6,7] The distance between JAERI and Kamioka is 295 km. Hyper-Kamiokande will enhance the capabilities of this experiment. In particular, if we wish to have a sensitivity to $\sin^2 2\theta_{\mu e}$ smaller than 0.01 in the $\nu_\mu \to \nu_e$ appearance experiment, an order of magnitude larger detector than SK will be needed. [6,7] Also, Konaka pointed out [8] that there is a possibility to measure CP violation in the neutrino sector using an intense low-energy ν_μ and $\bar{\nu}_\mu$ beams, provided that the solar-neutrino problem is solved by the MSW large mixing angle solution. For this experiment to be feasible for the CP phase down to 10 $\sim$ 20 degrees, Hyper-Kamiokande and intensity-upgraded (4MW beam power) JHF 50 GeV proton synchrotron will be needed. [8]

Hyper-Kamiokande may also be used as a far detector of neutrino factory experiments with a distance of several thousand kilometers. In this case, at least the charge of a muon produced in the detector volume must be identified with a very low mis-identification probability. It is not quite trivial to produce magnetic fields inside a very big water-Cherenkov detector volume and to operate photo-detectors in magnetic fields. However, this possibility is worth pursuing.

If Hyper-Kamiokande is sensitive to the $p \to \bar{\nu} K^+$ decay mode, prompt 6.3 MeV γ-ray emitted in a proton decay in oxygen should be measured so that the search for $p \to \bar{\nu} K^+$ can be made using the $K^+ \to \mu^+ \nu_\mu$ decay with prompt γ-ray tagging. [9] Then, the detection threshold is low enough to detect supernova neutrinos due to stellar collapse.[d] With this detector, we can observe an order of magnitude more neutrino events than SK. This enables us to study the mechanism of stellar collapse in detail, if it occurs in our Galaxy.

---

[c]The project including the construction of this accelerator has been known as the JHF (Japan Hadron Facility), and has been formally approved by the Japanese Government. It will be constructed by a collaborative effort of KEK and JAERI in the period of JFY (Japanese Fiscal Year) 2001 - 2006.

[d]Also, solar neutrinos may be observed with very high statistics and with reasonably low threshold, enabling us to study the day-night flux difference with much better accuracy than SK. Furthermore, a non-trivial seasonal variation, if any, of the solar-neutrino flux may be measured with high statistics. However, such measurements require fancy calibration systems over the entire fiducial volume. Careful assessment may be needed to evaluate if introducing such complexity to the next generation water Cherenkov detector pays, in particular, in view of the tradeoff between the cost and physics capabilities.

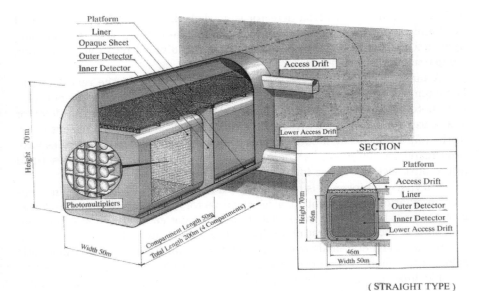

Figure 1: Possible configuration of Hyper-Kamiokande.

In particular, more than hundred electron neutrinos from neutronization burst will be detected and used to measure the ν_e mass down to less than 1 eV. With the 22.5 kton fiducial volume of SK, a stellar collapse in Andromeda cannot be detected through neutrino observations because only one neutrino event is expected on the average. Hyper-Kamiokande, with a fiducial volume of more than 10 times that of SK, can detect a bunch of neutrinos to signal a stellar collapse in Andromeda.

3 Toward the Design of Hyper-Kamiokande

Design of Hyper-Kamiokande is subject to a number of constraints. First of all, the site of the Kamioka Observatory is located ~ 1000 m underground, and excavation of a cavity for ~ 1 Mton water-Cherenkov detector needs careful feasibility study and assessment. Another important constraint is the cost. Cavity excavation and photo-detectors are the major factors determining the cost. Efforts are needed to reduce the cost, while keeping as much as possible the necessary capabilities for physics.

Figure 1 shows a possible configuration of Hyper-Kamiokande with a mass of 0.5 Mton. The water tank is a 200-m long tunnel with a cross-sectional area of 50 m × 50 m. If a 1 Mton detector is needed, the length of the tank should be doubled, or two 0.5 Mton detectors should be constructed. Considering light attenuation in water, the tank may be subdivided into four 50-m long compartments.

A number of R&D efforts are needed before designing the real detector. For example, simulations of the detector performance is needed to optimize the photo-detector configuration. Since the primary purpose of the detector is a search for nucleon decay, careful study should be done for the sensitivity to the important decay modes. In this direction, some simulation studies have been made for the $p \to e^+\pi^0$ decay mode, [10,11] and the results are described in the next section. Those for the $p \to \bar{\nu}K^+$ have yet to be done.

If Hyper-Kamiokande is to be used as a far detector of a long-baseline neutrino oscillation experiment at a neutrino factory, not only the muon charge measurement but also the measurement of the event energy and muon momentum will be necessary. First of all, feasibility studies are indispensable on how to produce magnetic fields over the detector volume. The effect of magnetic fields on the photo-detectors should also be simulated based on the test results.

The inner detector of SK is instrumented with Hamamatsu 20-inch photomultiplier tubes (PMTs). An important and interesting R&D item for Hyper-Kamiokande is development of new photo-detectors. Possibilities include obtaining a device with higher quantum efficiency, developing a flat and thin device, developing a device that can operate in high magnetic fields, etc.

Table 1: Three cases with different photocathode coverages were studied. The second column indicates the photocathode coverage relative to that of SK. The detection efficiency and background rate were obtained with the same or similar $p \to e^+\pi^0$ selection criteria as used in the analysis of SK. See text for the criteria.

Photocathode coverage	SK unit	Detection efficiency	Background per Mton·yr
40 %	1	43 %	~ 3
10 %	1/4	32 %	~ 3
4.4 %	1/9	21 %	~ 3

4 Simulation Studies for $p \to e^+\pi^0$

In this section we describe the results of a simulation studies made for the decay mode $p \to e^+\pi^0$. One of the purposes is to see how the sensitivity of a next-generation water Cherenkov detector to this decay mode depends on the photocathode coverage. (For photo-detectors, we assume 20-inch PMTs.)

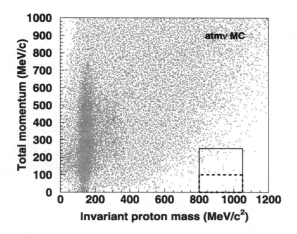

Figure 2: The total invariant mass and total momentum distribution of 20 Mton·yr atmospheric neutrino background. Assumed photocathode coverage is 40%. A window shown by the solid line is the same as the SK $p \to e^+\pi^0$ analysis. A tighter window shown by the dashed line gives a better signal-to-background ratio (see text).

As shown in Table 1, three cases were simulated. One case has the same photocathode coverage (40%) as SK. The other two cases have 10% (1/4 of SK) and 4.4% (1/9 of SK) photocathode coverage.[e] Generally, lower photocathode coverage results in a lower detection efficiency of proton decay and worse signal-to-background ratio. Another purpose is to study tighter criteria for the selection of $p \to e^+\pi^0$ candidates in order to improve the signal-to-background ratio. Specifically, a tighter total momentum cut was studied.

The simulation code used was that of the present SK. For the background study, 20 and 6.8 Mton·yr atmospheric neutrino events were generated for the case of 40% photocathode coverage and for the other two cases, respectively. Here, it should be noted that instead of simulating a larger-size detector, actually we simulated a detector having the same dimensions and the same optical properties (except photocathode caverage) as SK, with an exposure which is long enough to study the sensitivity for $p \to e^+\pi^0$ up to a partial lifetime of order 10^{35} years. In future, a simulation code has to be developed for a detector having the real dimensions. We, however, expect that the present simulation studies give reasonably accurate results for the assessment of the next-generation water Cherenkov detector as far as the $p \to e^+\pi^0$ decay mode

---

[e]To observe $p \to \bar{\nu}K^+$ by tagging a 6.3 MeV prompt γ, probably photocathode coverage cannot be less than $\sim 20\%$.

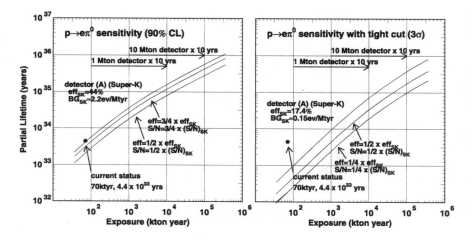

Figure 3: Left: 90% confidence level sensitivities for the partial lifetime of $p \to e^+\pi^0$ obtained with the same criteria as used in the present SK analysis. Right: 3σ sensitivities for the partial lifetime of $p \to e^+\pi^0$ obtained with a tighter P_{tot} cut. In both panels, the upper curve labeled "detector(A)(Super-K)" corresponds to the photocathode coverage of 40%, and the middle and lower curves correspond to the photocathode coverage of 10% and 4.4%, respectively. For the case of 40% photocathode coverage, the detection efficiency and background rate are the same as those of SK. Therefore, they are written as eff$_{\text{SK}}$ and BG$_{\text{SK}}$, respectively. S/N means the signal-to-background ratio.

is concerned.

We first studied the detection efficiency for $p \to e^+\pi^0$ and the background rate by adopting the same or similar criteria as the SK analysis[12] for $p \to e^+\pi^0$. For the case of 40% photocathode coverage, the selection criteria are (i) there must be 2 or 3 Cherenkov rings; (ii) all rings must be showering; (iii) for the case of 3 rings, $85 < M_{\pi^0} < 185$ MeV/c^2; (iv) there must be no decay electron; (v) $800 < M_p < 1050$ MeV/c^2 and $P_{\text{tot}} < 250$ MeV/c, where M_{π^0} is the reconstructed π^0 mass, M_p is the invariant proton mass, and P_{tot} is the total momentum. For the other two cases, criteria (ii) and (iii) were not imposed. Moreover, for the case of 4.4% photocathode coverage, the total invariant mass cut was loosened as $750 < M_p < 1050$ MeV/c^2.[10,11]

The detection efficiencies were obtained from the $p \to e^+\pi^0$ simulation, and are listed in the third column of Table 1. Figure 2 shows the total invariant mass and total momentum distribution of the simulated 20 Mton·yr atmospheric neutrino events which are selected with criteria (i) $\sim$ (iv). The assumed photocathode coverage is 40%. There are 45 events in the signal window (criterion (v)). Therefore, the background rate is $\sim$ 3 events per Mton·yr.

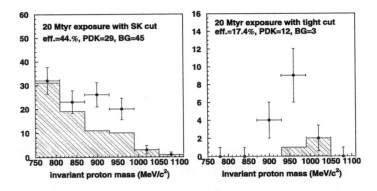

Figure 4: Left: Proton invariant mass distribution for the case of SK cuts. Right: Proton invariant mass distribution for the case of a tighter P_{tot} cut. The hatched histograms show the atmospheric neutrino background. In both cases, the partial lifetime of $p \to e^+\pi^0$ of 10^{35} years, an exposure of 20 Mton·yr, and the detector photocathode coverage of 40% are assumed.

For the other two cases, the background rate also turned out to be ~ 3 events per Mton·yr.

The left panel of Fig. 3 shows the resulting 90% confidence level sensitivities. It is seen that the sensitivity to $\tau_p/B(p \to e^+\pi^0) = 10^{35}$ years can be reached with exposure of ~ 5 Mton·yr for the case of 40% photocathode coverage, and that of ~ 8 Mton·yr for the case of 10% photocathode coverage. Assuming $\tau_p/B(p \to e^+\pi^0) = 10^{35}$ years, 29 proton decay events would be observed[f] with 45 atmospheric neutrino events for an exposure of 20 Mton·yr. The left panel of Fig. 4 shows the corresponding proton invariant mass distribution. As can be seen, the proton decay signal is not very clear. The signal can be extracted only when the atmospheric neutrino background is well understood.

To improve the signal-to-background ratio, a tighter cut on the total momentum is very effective. By imposing $P_{\text{tot}} < 100$ MeV/c, background events from 20 Mton·yr Monte Carlo sample reduce to 3 events (see Fig.2) for the case of 40% photocathode coverage. The detection efficiency for $p \to e^+\pi^0$ is 17.4%. Assuming $\tau_p/B(p \to e^+\pi^0) = 10^{35}$ years, 12 proton decay events would be observed[f] for an exposure of 20 Mton·yr. The $P_{\text{tot}} < 100$ MeV/c cut highlights the decay of free protons as can be seen from Fig. 5. Consequently, the proton mass peak can be seen in the total invariant mass distribution (right panel of

---

[f] The numbers of proton decay events presented at the Workshop were found to be wrong. They should be reduced by a factor of 10/18.

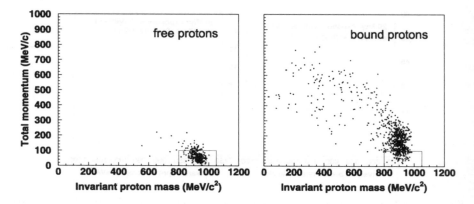

Figure 5: The total invariant mass and total momentum distributions for free protons (left) and for protons bound in the ^{16}O nuclei (right). The events were generated according to the ratio, free:bound = 2:8.

Fig. 4). The signal-to-background ratio is 4 for $\tau_p/B(p \to e^+\pi^0) = 10^{35}$ years. Therefore, it is better than 1 for $\tau_p/B(p \to e^+\pi^0) < 4 \times 10^{35}$ years. This suggests the possibility of discovering $p \to e^+\pi^0$ decay if the partial lifetime is shorter than a few $\times 10^{35}$ years.

With the tight total momentum cut, we now consider 3σ sensitivities for $p \to e^+\pi^0$. The upper curve in the right panel of Fig. 3 corresponds to the case of 40% photocathode coverage. For the cases of 10% and 4.4% photocathode coverages, efficiencies were studied, but the background rates were assumed to be the same as the case of 40% photocathode coverage: the results are also shown in the same figure. Assuming $\tau_p/B(p \to e^+\pi^0) = 10^{35}$ years, exposures of 15 Mton·yr and 40 Mton·yr are needed for 40% and 10% photocathode coverages, respectively, to obtain the 3σ significance of the proton decay signal.

As far as the $p \to e^+\pi^0$ decay with a partial lifetime of order 10^{35} years is concerned, the simulation studies suggest that with a ~ 1 Mton water Cherenkov detector having photocathode coverage of 10% $\sim$ 40%, one can observe the proton decay signal within a reasonable time range.

5 Conclusions

For a next-generation nucleon-decay experiment, a water-Cherenkov detector with total mass of ~ 1 Mton is considered. Assuming that such a detector is constructed at the site of the Kamioka Observatory, it is called Hyper-Kamiokande. It should be sensitive not only to the gauge-boson-mediated decay mode $p \to e^+ \pi^0$, but also to the dominant decay modes predicted by SUSY Grand Unified Theories, i.e., decay modes involving K mesons. Then, the detector has a reasonably low threshold, and it will allow us to observe an order of magnitude more stellar-collapse neutrinos than Super-Kamiokande. Hyper-Kamiokande will be used as a far detector in the Tokai-to-Kamioka long baseline neutrino oscillation experiment with a high-intensity 50 GeV proton synchrotron to be constructed at JAERI as a neutrino source. It is also worth studying if this detector can be used as a far detector of a long base-line neutrino oscillation experiment at a neutrino factory.

Simulation studies of the $p \to e^+ \pi^0$ decay suggest that with a ~ 1 Mton water Cherenkov detector having photocathode coverage of 10% $\sim$ 40%, one can observe the proton decay signal within a reasonable time range.

References

1. The Super-Kamiokande Collaboration, Y. Fukuda *et al.*, Phys. Rev. Lett. **81** (1998) 1562.
2. M. Shiozawa, talk presented at the 30th Int. Conf. on High Energy Physics (ICHEP2000), 2000, Osaka, Japan.
3. W. Marciano, in these Proceedings.
4. K. Nakamura, talk presented at the Workshop on the Next Generation Nucleon Decay and Neutrino Detector, 1999, SUNY at Stony Brook.
5. K. Nakamura, in *Neutrino Oscillations and Their Origin*, edited by Y. Suzuki, M. Nakahata, M. Shiozawa, and K. Kaneyuki (Universal Academy Press, Tokyo, 2000), p. 359.
6. JHF Neutrino Working Group, Y. Itow *et al.*, *Letter of Intent: A Long Baseline Neutrino Oscillation Experiment Using the JHF 50 GeV Proton-Synchrotron and the Super-Kamiokande Detector*, February 2000.
7. Y. Obayashi, in these Proceedings.
8. A. Konaka, private communication.
9. The Super-Kamiokande Collaboration, Y. Hayato *et al.*, Phys. Rev. Lett. **83** (1999) 1529.
10. M. Shiozawa, in *Next Generation Nucleon Decay and Neutrino Detector (NNN99)*, edited by M.V. Diwan and C.K. Jung (AIP Conference Proceedings 533, AIP, New York, 2000) p. 21.
11. M. Shiozawa, in *Neutrino Oscillations and Their Origin*, edited by Y. Suzuki, M. Nakahata, M. Shiozawa, and K. Kaneyuki (Universal Academy Press, Tokyo, 2000), p. 365.
12. The Super-Kamiokande Collaboration, M. Shiozawa *et al.*, Phys. Rev. Lett. **81** (1998) 3319.

A WATER ČERENKOV CALORIMETER AS THE NEXT GENERATION NEUTRINO DETECTOR

YI-FANG WANG

Stanford University, Department of Physics, Stanford, CA 94305, USA
E-mail: yfwang@hep.stanford.edu

We propose here a large homogeneous calorimeter as the next generation neutrino detector for ν factories and/or conventional ν beams. The active media is chosen to be water for obvious economical reasons. The Čerenkov light produced in water is sufficient to have good energy resolution, and the pattern recognition is realized by a modular water tank structure. Monte Carlo simulations demonstrate that the detector performance is excellent for identifying neutrino CC events while rejecting background events.

1 Introduction

Neutrino factories and conventional beams have been discussed extensively in the literature[1] as the main facility of neutrino physics for the next decade. The main physics objectives include the measurements of $sin\theta_{13}$, Δm^2_{13}, the leptonic CP phase δ and the sign of Δm^2_{23}. All of these quantities can be obtained through the disappearance probability $P(\nu_\mu \to \nu_\mu)$ and the appearance probability $P(\nu_\mu(\nu_e) \to \nu_e(\nu_\mu))$ and $P(\bar\nu_\mu(\bar\nu_e) \to \bar\nu_e(\bar\nu_\mu))$. To measure these quantities, a detector should: 1) be able to identify leptons: e, μ and if possible τ; 2) have good pattern recognition capabilities for background rejection; 3) have good energy resolution for event selection and to determine $P_{\alpha\to\beta}(E)$; 4) be able to measure the charge for $\mu^\pm$ in the case of ν factories; and 5) be able to have a large mass(100-1000 kt) at an affordable price.

	Iron Calorimeter	Liquid Ar TPC	Water Ring Imaging	Under Water/Ice Čerenkov counter
Mass	10-50 kt	1-10 kt	50-1000 kt	100 Mt
Charge ID	Yes	Yes	?	No
E resolution	good	very good	very good	poor
Examples	Minos Monolith	ICANOE	Super-K, Uno Aqua-rich	Amanda, Icecube Nestor, Antares

Table 1. Currently proposed detector for ν factories and conventional ν beams.

Currently there are four types of detectors proposed[1,2], as listed in table 1. These detectors are either too expensive to be very large, or too large to have a magnet for charge identification. In this talk, I propose a new type of detector – a water Čerenkov calorimeter – which fulfills all the above requirements.

2 Water Čerenkov Calorimeter

Water Čerenkov ring image detectors have been successfully employed in large scale, for obvious economic reasons, by the IMB and the Super-Kamiokande experiments. However a substantial growth in size beyond these detectors appears problematic because of the cost of excavation and photon detection. To overcome these problems, we propose here a water Čerenkov calorimeter with a modular structure, as shown in Fig. 1.

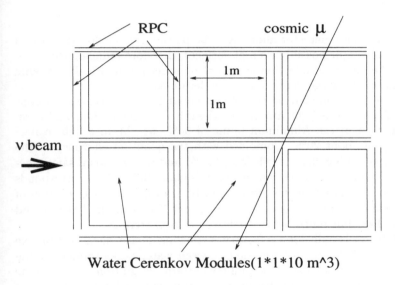

Figure 1. Schematic of water Čerenkov calorimeter

Each tank has dimensions $1 \times 1 \times 10 m^3$, holding a total of 10 t of water. The exact segmentation of water tanks is to be optimized based on the neutrino beam energy, the experimental hall, the cost, etc. For simplicity, we discuss in the following 1 m thick tank, corresponding to 2.77 X_0 and 1.5 λ_0. The water tank is made of PVC with Aluminum lining. Čerenkov light is reflected by Aluminum and transported towards the two ends of the tank, which are covered by wavelength shifter(WLS) plates. Light from the WLS is guided to a 5" photon-multiplier tube(PMT), as shown in Fig. 2. The modu-

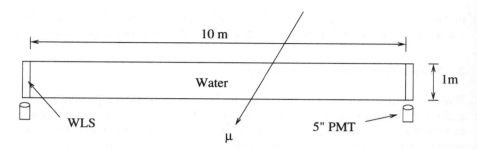

Figure 2. Schematic of a water tank.

lar structure of such a detector allows it to be placed at a shallow depth in a cavern of any shape(or possibly even at surface), therefore reducing the excavation cost. The photon collection area is also reduced dramatically, making it possible to build a large detector at a moderate cost.

A through-going charged particle emits about 20,000 Čerenkov photons per meter. Assuming a light attenuation length in water of 20m and a reflection coefficient of the Aluminum lining of 90%, we obtain a light collection efficiency of about 20%. Combined with the quantum efficiency of the PMT(20%), the WLS collection efficiency(25%) and an additional safety factor of 50%, the total light collection efficiency is about 0.5%. This corresponds to 100 photoelectrons per meter, which can be translated to a resolution of $4.5\%/\sqrt{E}$. This is slightly worse than the Super-Kamiokande detector and liquid Argon TPC but much better than iron calorimeters[1].

If this detector is built for a ν factory, a tracking device, such as Resistive Plate Chambers (RPC)[3] will be needed between water tanks to identify the sign of charge. RPCs can also be helpful for pattern recognition, to determine precisely muon directions, and to identify cosmic-muons for either veto or calibration. The RPC strips will run in both X- and Y-directions with a width of 4 cm. A total of $\sim 10^5$ m^2 is needed for a 100 kt detector, which is more than an order of magnitude larger than the current scale[3]. R&D efforts would be needed to reduce costs.

The magnet system for such a detector can be segmented in order to minimize dead materials between water tanks. If the desired minimum muon momentum is 5 GeV/c, the magnet must be segmented every 20 m. Detailed magnet design still needs to be worked out; here we just present a preliminary idea to start the discussion. A toroid magnet similar to that of Minos, as shown in Fig. 3, can produce a magnetic field B > 1.5 T, for a current

$I > 10^4$ A. The thickness of the magnet needed is determined by the error from the multiple scattering: $\Delta P/P = 0.0136\sqrt{X/X_0}/0.3BL$, where L is the thickness of magnet. For L=50 cm, we obtain an error of 32%. The measurement error is given by $\Delta P/P \simeq \delta\alpha/\alpha = \sigma P/0.3rBL$, where r is the track length before or after the magnet and σ is the pitch size of the RPC. For P=5 GeV/c, $\sigma = 4$ cm and r=10 m, the measurement error is 9%, much smaller than that from multiple scattering. It should be noted that P_μ is also measured from the range. By requiring that both P_μ measurements are consistent, we can eliminate most of the fake wrong sign muons. The iron needed for such a magnet is about 20% of the total mass of the water.

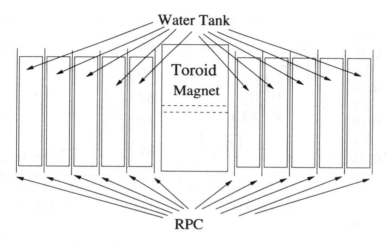

Figure 3. Schematic of a toroid magnet

The cost of such a detector is moderate compared to other types of detectors, enabling us to build a detector as large as 100 - 1000 kt. The combination of size, excellent energy resolution and pattern recognition capabilities makes this detector very attractive. An incomplete but rich physics program can be listed as follows: 1) neutrino physics from ν factories or ν beams; 2) improved measurements of atmospheric neutrinos; 3) observation of supernovae at distances up to hundreds of kpc; 4) determination of primary cosmic-ray composition by measuring multiple muons; 5) searches for WIMP's looking at muons from the core of the earth or the sun with a sensitivity covering DAMA's allowed region; 6) searches for monopoles looking at slow moving particles with high dE/dx; 7) searches for muons from point sources; 8) searches for exotic particles such as fractionally charged particles.

Depending on the location of the detector, other topics on cosmic-ray physics can be explored.

3 Performance of Water Čerenkov Calorimeter

To study the performance of such a detector, we consider in the following two possible applications in the near future: JHF neutrino beam to Beijing with a baseline of 2100 km and NuMi beam from Fermilab to Minos with a baseline of 735 km. The energy spectra of visible ν_μ CC events are shown in Fig. 4.

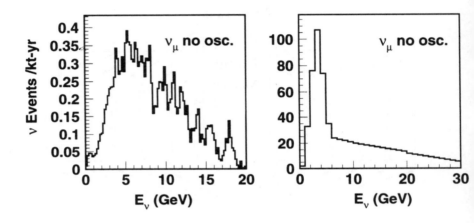

Figure 4. Beam profile of JHF-Beijing and Numi-Minos.

We use a full GEANT Monte Carlo simulation program and the Minos neutrino event generator. A CC ν signal event is identified by its accompanying lepton, reconstructed as a jet. Fig. 5 shows the jet energy normalized by the energy of the lepton. It can be seen from the plot that leptons from CC events can indeed be identified and the jet reconstruction algorithm works properly. It is also shown in the figure that the energy resolution of the neutrino CC events is about 10% in both cases.

The neutrino CC events are identified by the following 5 variables: E_{max}/E_{jet}, L_{shower}/E_{jet}, N_{tank}/E_{jet}, R_{xy}/E_{tot}, and R_{xy}^{max}/E_{tot}, where E_{jet} is the jet energy, E_{tot} the total visible energy, E_{max} the maximum energy in a cell, L_{shower} the longitudinal length of the jet, N_{tank} the number of cells with energy more than 10 MeV, R_{xy} the transverse event size and R_{xy}^{max} the

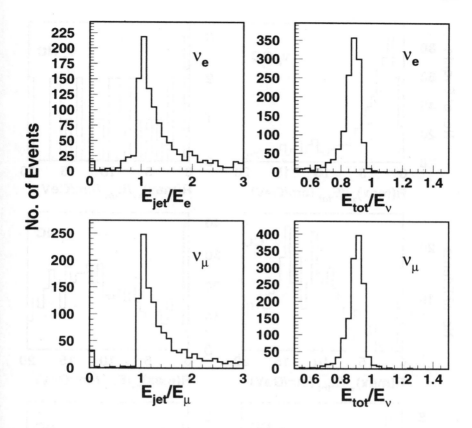

Figure 5. The reconstructed jet energy and total visible energy. The fact that E_{jet}/E_{lepton} peaks around one shows that the jet reconstruction algorithm finds the lepton from CC events. The fraction of total visible energy to the neutrino energy indicates that we have an energy resolution better than 10% for all neutrinos. The bias is due to invisible neutral hadrons and charged particles below Čerenkov thresholds.

transverse event size at the shower maxima. Fig. 6 shows R_{xy}^{max}/E_{tot} for all different neutrino flavors. It can be seen that ν_e CC events can be selected with reasonable efficiency and moderate backgrounds. Table 2 shows the final results from this pilot Monte Carlo study. For ν_e and ν_μ events, ν_τ CC events are dominant backgrounds, while for ν_τ, the main background is ν_e. It is interesting to see that this detector can identify ν_τ in a statistical way. Similar results are obtained for a detector with 0.5m water tanks without RPCs.

248

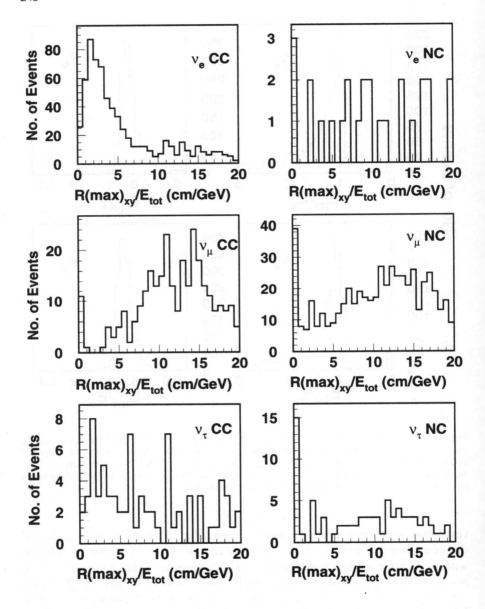

Figure 6. The transverse event size at the shower maxima for various type of neutrino events. The distribution of ν_e is different from all the others.

These results are similar to or better than those from water Čerenkov image detectors[4] and iron calorimeters[5].

	JHF-Beijing			NuMi-Minos	
	ν_e	ν_μ	ν_τ	ν_e	ν_μ
CC Eff.	30%	53%	9.3%	15%	53%
ν_e CC	-	>1300:1	3:1	-	>1300:1
ν_e NC	166:1	665:1	60:1	600:1	>610:1
ν_μ CC	700:1	-	270:1	14000:1	-
ν_μ NC	92:1	>6000:1	39:1	320:1	2000:1
ν_τ CC	20:1	12:1	-	33:1	18:1
ν_τ NC	205:1	1100:1	61:1	530:1	3200:1

Table 2. Results from Monte Carlo simulation: Efficiency vs background rejection power for different flavors.

4 Summary

In summary, the water Čerenkov calorimeter is a cheap and effective detector for ν factories and ν beams. The performance is excellent for ν_e and ν_τ appearance and ν_μ disappearance from a Monte Carlo simulation. Such a detector is also very desirable for cosmic-ray physics and astrophysics. There are no major technical difficulties although R&D and detector optimization are needed.

Acknowledgments

I would like to thank G. Gratta, S. Wojcicki, L. Wai and H.S. Chen for many useful discussions.

References

1. See for example, C. Albright et al., hep-ph/0008064.
2. K. Dick et al, hep-ph/0008016.
3. C. Bacci et al., Nucl. Phys. Proc. Suppl. 78 (1999) 38.
4. Y. Itow et al., "Letter of Intent: A Long Baseline Neutrino Oscillation Experiment using JHF 50 GeV proton-Synchrotron and the Super-Kamiokande Detector".
5. L. Wai, private communication.

LFV2000 Scientific Program

Monday, October 2, 2000

Time	Topic	Speaker
9:00 ~ 9:10	Welcome	S. Olsen
9:10 ~ 10:00	Muon Applied Science	K. Nagamine
10:20 ~ 11:10	Proton Decay and Neutrino Mass	K.S. Babu
11:10 ~ 12:00	Theoretical Motivation for Lepton Flavor Violation (LFV)	J. Feng
13:30 ~ 14:20	SUSY GUT and LFV	Y. Okada
14:20 ~ 15:00	LFV and SUSY with Right-handed Neutrino	D. Nomura
15:00 ~ 15:40	LFV and R Parity Violating Models	K. Tobe
16:00 ~ 16:40	LFV and Future Lepton Colliders	M. Tanaka
16:40 ~ 17:20	LFV Studies with K Decays	W. Molzon

Tuesday, October 3, 2000

Time	Topic	Speaker
9:00 ~ 9:50	Neutrino Oscillation Scenarios and GUT Model Predictions	C. Albright
9:50 ~ 10:40	Neutrino Mass and Extra Dimensions	R.N.Mohapatra
11:00 ~ 11:40	$\mu \to$ e Conversion Phenomenology	A. Czarnecki
11:40 ~ 12:20	$\mu \to$ e Conversion: An Interplay between Particle, Nuclear and Non & Standard Physics	T.S. Kosmas
13:30 ~ 14:10	Status of $\mu \to$ e Conversion at PSI	P. Wintz
14:10 ~ 14:45	MECO (1) - Experiment -	J. Sculli
14:45 ~ 15:20	MECO (2) - Beam -	V. Tumakov
15:40 ~ 16:15	PRISM (1) - Beam -	Y. Kuno
16:15 ~ 16:50	PRISM (2) - Experiment -	M. Aoki
16:50 ~ 17:30	$\mu \to$ e γ Experiment at PSI	J. Yashima

Wednesday, October 4, 2000

Time	Topic	Speaker
9:00 ~ 9:50	Leptogenesis	E. Ma
	<Session of Atmospheric Neutrino>	
9:50 ~ 10:30	Neutrino Cross Sections	D. Casper
10:50 ~ 11:30	Calculation of Atmospheric Neutrino Flux	M. Honda
11:30 ~ 12:10	CP Violation and Atmospheric Neutrinos	I. Stancu

Thursday, October 5, 2000

Time	Topic	Speaker
9:00 ~ 9:50	Neutrino Factory Phenomenology	V. Barger
9:50 ~ 10:30	Four-generation Neutrino Oscillation	O. Yasuda
10:50 ~ 11:30	Neutrino Oscillations and New Physics	L. Johnson
11:30 ~ 12:10	CP violation in Neutrino Oscillation at Low energy	J. Sato
	<Sesson of Neutrino Factory Plans>	
13:30 ~ 14:20	US Neutrino Factory Studies	R. Palmer
14:20 ~ 15:10	Neutrino Factory Studies at CERN	A. Riche
15:30 ~ 16:20	Neutrino Factory Studies in Japan	Y. Mori

Friday, October 6, 2000

Time	Topic	Speaker
9:00 ~ 9:50	CPT Violation in Neutrino Oscillation	S. Pakvasa
	<Session on Comparison of Conventional Beams vs. Neutrino Factory>	
9:50 ~ 10:30	TBA	J. Learned
10:50 ~ 11:30	Conventional Beam at JHF	Y. Obayashi
11:30 ~ 12:10	Molybdenum Observatory of Neutrinos (MOON) and Intense Neutrino Source	H. Ejiri
	<Session on Large Detectors for Neutrino and Nucleon Decay>	
13:30 ~ 14:10	Future Nucleon Decay Detector at Kamioka	K. Nakamura
14:10 ~ 14:50	UNO Detector	D. Casper
14:50 ~15:30	A Water Cerenkov Calorimeter for the Next Generation Neutrino Detector	Y. Wang
16:00 ~ 17:00	Summary	W. Marciano